Selected Titles in This Series

The Game's Afoot!
Game Theory in
Myth and Paradox

Alexander Mehlmann

Game-Theoretic Rebus

chessboard	the world
tortoise	paradox
key	strategy
dice	chance
sword	conflict
coin	necessity
currants	conclusion

STUDENT MATHEMATICAL LIBRARY
Volume 5

The Game's Afoot! Game Theory in Myth and Paradox

Alexander Mehlmann

Translated by
David Kramer

AMERICAN MATHEMATICAL SOCIETY

Editorial Board

Originally published in German by Friedr. Vieweg & Sohn
Verlagsgesellschaft mbH, D-65189 Wiesbaden, Germany, under the title
"Wer gewinnt das Spiel? Spieltheorie in Fabeln und Paradoxa. 1. Auflage
(1st edition)" by Alexander Mehlmann. © by Friedr. Vieweg & Sohn
Verlagsgesellschaft mbH, Braunschweig/Wiesbaden, 1997.

2000 *Mathematics Subject Classification*. Primary 91-01, 91A05, 91A40,
97A20, 97A90.

Library of Congress Cataloging-in-Publication Data

Mehlmann, Alexander.
   [Wer gewinnt das Spiel. English]
   The game's afoot! : game theory in myth and paradox / Alexander Mehlmann.
      p. cm. — (Student mathematical library, ISSN 1520-9121 ; v. 5)
   Includes bibliographical references and index.
   ISBN 0-8218-2121-0 (alk. paper)
   1. Game theory.   I. Title.   II. Series.
QA269.M44513   2000
519.3—dc21
                                               00-020915

For **Grace**, my best-beloved adversary,
and **Sabrina**, our mutual game value

Contents

Contents

Foreword

> Enigmatic for the masses,
> playfully with life we fool.
> That which human wits surpasses
> draws our special ridicule.[1]
> —**Christian Morgenstern,** *Gallows Hill*

No other mathematical discipline has altered the study of economics, the social sciences, and biology as has game theory, in the fifty years since its inception. Social traps, political mock battles, evolutionary confrontations, economic struggles, and not least literary conflicts can all be viewed as "games" of this theory.

This book is addressed to readers who are prepared to consider the perspective of a study, both formal and with practical application, that embraces science, literature, and life's conflicts.

For the layperson unencumbered by any previous knowledge of game theory, an introduction to the subject does not require graduate study in higher mathematics. An ability to think logically that does not shrink from entertaining sophistry will do nicely for passing unscathed through the hall of mirrors of strategic decision-making.

With the help of formulas, fables, and paradoxes we shall begin our lighthearted excursion into the world of strategic calculation. The

---

[1]Call it infantile vendetta/ on life's deeply serious aim—/ you will know existence better/ once you understand our game. [Translation by Max Knight.]

stations of this journey support the mathematics of conflict, and provide a connecting thread through the labyrinth of solution concepts and the unraveling of the myths of game theory. Our fanciful introduction to contemporary mathematical game theory stretches from the dilemma of the arms race by way of disaster on the internet to a lesson in the just division of a cake.

If there is a model for this undertaking, then it must be the book that made accessible to me—during my far-off student days—the notions of game, strategy, and saddle point, namely, J.D. Williams's *The Compleat Strategyst: Being a Primer on the Theory of Games of Strategy* [**95**].

Every refreshing inclination that winked at me from this collection incited my appetite for game-theoretic excursions in literary realms. I invite the reader who is so disposed to follow me on this path to an appreciation of game theory.

Acknowledgments

He thought he saw the Unicorn, the Virgin's wildest pet,
He looked again and saw it was a Long Outstanding Debt.
He wrote and wrote and wrote and wrote—and hasn't written yet.

—**G. K. Chesterton.** *An apology for not writing*

To write a book is surely a game of patience, indeed sometimes a most dangerous game. But it is never entirely a one-person game. I offer here my thanks to human and institutional teammates:

- To **Bruno Niederle** and **Heinrich Vierhapper**, whose considerable medical skill and empathy brought me in good health once again into the game.

- To **Erich Lessing** for his masterly pictures to the *Odyssey* [**55**], which transport the viewer into the golden realms of ancient myth.

- To **Ulrike Schmickler-Hirzebruch**, of the editorial department of Vieweg publishers, for the gentle pressure that accompanied me through all the phases of the creation of the manuscript.

- To **Maria Dworak** and **Georg Giokas** (as representatives for all the participants in my course on game theory) for their alert curiosity, without which no player can achieve success.

- To **Blue Sky Research**[2] for their magical product Textures 1.8, which made possible a classic impression of the manuscript of this book.[3]

- To **Dov Samet**, the nonpareil publisher of the *International Journal of Game Theory*, for an hour of instruction by correspondence in game-theoretic mythology.

- To **Barbara Boock-James**, of the library of the University of Freiburg, for guiding me across the internet to Uhland's "Der weisse Hirsch."

- To **Gernot Tragler** for the professional photographs of my model of the game-theoretic rebus.

- To **David Kramer** for his skills in translating both mathematical and literary passages.

---

[2]My favorite software company in distant Portland, Oregon: 317 SW Alder, Portland, OR 97204.

[3]Translator's note: The translation was produced with the help of the even more magical Textures 2.0.

List of Figures

Part I

The Glass Bead Game

Introduction to Part I

Wir lassen vom Geheimnis uns erheben
Der magischen Formelschrift, in deren Bann
Das Uferlose, Stürmende, das Leben,
Zu klaren Gleichnissen gerann.[4]
—**Hermann Hesse,** *Das Glasperlenspiel*

When Hermann Hesse's *The Glass Bead Game* [**40**] appeared in 1943, anyone who had ventured the opinion that this novel's story of a game of the intellect could be taken as anything more than a literary fiction would have been roundly ridiculed. Yet only a few months later the birth of game theory had translated the coordinates of human knowledge, and the strange parallels between literary and scientific imagination were already discernible.

---

[4]We draw upon the iconography/ Whose mystery is able to contain/ The boundlessness, the storm of all existence,/ Give chaos form, and hold our lives in rein. [*The Glass Bead Game*, translation by Richard and Clara Winston.]

The tool used by John von Neumann[5] and Oskar Morgenstern[6] in their seminal monograph *Theory of Games and Economic Behavior* [**71**] to study the values of mankind was indisputably mathematics. And yet, if one follows the history of this glass bead game back through the centuries, one encounters in every nook and cranny a vast treasure trove of motifs and situations that—with contributions from the most diverse disciplines—have significantly influenced the development of game theory.

Traditional historiographers are wont to trace the gnarled roots of game-theoretic argumentation only as far back as the gambling dens of the late Renaissance or early Baroque. From the viewpoint of aleatoric and combinatorial game theory this seems to be a thoroughly reasonable and correct approach.

The authors of the first scientific works on games of chance are counted among mathematicians of the first rank. Girolamo Cardano and Galileo devoted their attentions to chance and their eyes to games of dice. Blaise Pascal and Pierre de Fermat commented in their correspondence on the fundamental problems of gambling and just payoffs raised by the professional gambler Chevalier de Méré. With Christiaan Huygens [**42**] we have finally reached the starting point of a development at whose terminus stands today's theory of probability.

Already in the year 1612 Bachet de Méziriac [**63**] had calculated the winning positions in a simple combinatorial game. Two players alternately add a number between 1 and 10 to the running subtotal. The game begins at 0 and ends in victory for the player who first reaches the total 100. For the most general form of this class of problems, the so-called nim games, it was shown by E. H. Moore in 1909 [**64**] that under certain circumstances it is more blessed to receive than to give. From these first modest efforts has arisen the gleaming apparatus of combinatorial game theory.

---

[5]The mathematician **John** (**Johann**, also **János**) **von Neumann** (born 1903 in Budapest, died 1957 in Washington, D.C.) worked, after completing his studies in the science of chemical processes (Zürich) and mathematics (Budapest), as a lecturer in Göttingen, Berlin, and Hamburg, and from 1933 at Princeton University in the Institute for Advanced Studies.

[6]**Oskar Morgenstern** (born 1902 in Görlitz, died 1977 in Princeton) worked until 1938 as a professor of economics in Vienna, thereafter at Princeton University (until 1970) and New York.

In 1713 J. Waldegrave [93] analyzed a card-exchange problem in the card game "Le Her," the solution of which could be described in terms of a certain random mechanism in the choice of strategy. These findings were forgotten; upon their rediscovery in the 1960s they were recognized as the first appearance of an example of a mixed minimax strategy in an antagonistic strategic game. This new category was meanwhile defined by numerous examples in several works [10, 11] of Emile Borel in the years 1921–1927. Independently, von Neumann proved the minimax theorem in full generality in 1928 [70]. All in all, however, these beginnings left little trace.

It remained to von Neumann and Morgenstern to inaugurate a sustained era of game theory in their above-mentioned monograph. The programmatic goal, as is clear from the title, was not so much the application of the new theory to games as to the broad field of economics and social problems. It may come as a surprise to modern dogmatic game theorists, but both authors could as well be associated with decision situations of a lighthearted literary provenance, as it finds expression, for example, in the analysis of the confrontation between Sherlock Holmes and his eternal adversary Professor Moriarty.[7]

In Part I of this book we shall attempt to make available to the interested reader a clear, informal, and yet formal—but only to the extent necessary—look back on the first fifty years of game theory. The understanding of definitions, solution strategies, and methods of proof as well, will be assisted by fables, riddles, and paradoxes, which will make possible deep insights into the nature of strategic thinking.

Grossly undervalued by the official chronicles of game theory, such contributions, in the role of fellow travelers of this new discipline, have anticipated current motifs and not least have in a lighthearted way made more permeable the rigid boundaries of a thoroughly mathematical discourse. Thus it is not surprising that behind Selten's

---

[7]The Napoleon of Crime, Professor of mathematics at provincial universities, author of—from a lack of suitable experts on the subject—the never discussed, yet considered a masterwork, *Dynamics of an Asteroid*. See Arthur Conan Doyle's "The Final Problem" in [18] and Section 7.1 for a more complete description of the basic pursuit game.

(economic) chain store model [86] lies nothing other than the amusing paradox of Quine about the hanged man (see [32, 41]). Indeed, it was in the form of an anecdote told by A. W. Tucker—as the idea for the "prisoner's dilemma"—that the experimental game originally developed by Flood and Dresher became the hackneyed synonym for social and thermonuclear traps.

Indeed, there is no—to paraphrase the immortal words that Euclid uttered to Ptolemy Soter (or was it Menæchmus to Alexander the Great? There seems to be disagreement on this point)—royal road to the mathematics of conflicts. The interested reader must nonetheless have no fear of being impaled on a tabular display or of being unable to see the apparently impenetrable forest of specialist publications on account of the large number of game trees. The most important goal of our efforts consists in interpreting the fascinating myths of game theory and weaving an Ariadne's thread through the labyrinth of solution concepts.

Chapter 1

Games, Form(ula)s, and Scholars

> Gelegentlich ergreifen wir die Feder
> und schreiben Zeichen auf ein weisses Blatt,
> Die sagen dies and das, es kennt sie jeder,
> Es ist ein Spiel, das seine Regeln hat.[1]
> —**Hermann Hesse.** *Das Glasperlenspiel*

The simplest model for representing a noncooperative game, its *normal*, or *strategic*, *form*,[2] presupposes three conditions:

Normal Form Representation

1. *Specification of the players.*

2. *Complete description of the strategies available to each player.*

3. *Specification of the payoff[3] values that accrue to the various players for every possible strategic constellation.*

---

[1] From time to time we take our pen in hand/ And scribble symbols on a blank white sheet./ Their meaning is at everyone's command;/ It is a game whose rules are nice and neat. [*The Glass Bead Game*, translation by Richard and Clara Winston.].

[2] While in a noncooperative game the individual and his strategic decisions stand in the foreground, the leitmotif of a cooperative game is the question of the benefit that a group (or *coalition*) of players can achieve in common, in addition to the rules of a just division of what is won (or lost) among the members of the coalition.

[3] Here we come to some classical party-game jargon. In a more neutral tone, rather than "payoff" one would employ the term *utility value*.

1.1. Scissors–Stone–Paper

In the well-known children's game scissors–stone–paper two players each simultaneously display a hand in one of three configurations. A flat hand represents a sheet of paper, a fist indicates a stone, and an extended index and middle finger stand for a pair of scissors. The possible outcomes of the game are given by the following rule: Scissors cuts paper; paper wraps stone; stone breaks scissors.

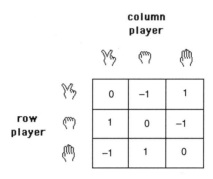

Figure 1.1. Scissors–stone–paper

The normal form of the scissors–stone–paper game can be given by a simple table, as is apparent from Figure 1.1, whose payoff values for victory (1), defeat (−1), and a draw (0) are shown from the point of view of the first player (the "row" player). Such a table is also known as the *game matrix* $A = (a_{ij})$ of the game.

Scissors–stone–paper belongs to the category of two-person zero-sum games. Whatever is won by one player is simultaneously lost by the other. If we let $B = (b_{ij})$ denote the payoff matrix of the second player, then in a two-person zero-sum game the relation $B = -A$ always holds.

Scissors–stone–paper can also be identified as a symmetric two-person game. In this class of games the payoff matrix B of the second player can be obtained from the matrix A of the first player by a simple reflection along the main diagonal; that is, for every choice of a row (strategy, action) i made by the first player and of a column (strategy, action) j made by the second player the relation $b_{ij} =$

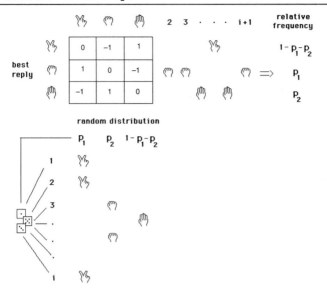

Figure 1.2. One-sided, shortsighted play

a_{ji} holds. This state of affairs is indicated in matrix notation by the expression $B = A^{\mathrm{t}}$, where A^{t} denotes the transpose of A, the matrix formed by interchanging the (i, j) and (j, i) entries of A for all pairs i, j. In a symmetric game, interchanging the roles of the two protagonists has no effect on the game.

An intuitive approach to solving scissors–stone–paper begins with the following considerations.

The row player has a unique *best reply* for every action of the column player (for example, always to choose "scissors" when the opponent chooses "paper"). He can thereby always emerge the victor, provided, of course, that he knows in advance the choice his opponent will make. Therefore, the column player is compelled to conceal his actions. In a game whose attraction is in repetition, such a concealment can be attained by the player's choosing at random from among the various possibilities available to him. The rules of the game are such that the manner in which the players let chance decide their moves must be apparent to all players.

The column player can now wield a pair of scissors with probability p_1 and a stone with probability p_2. If this *mixed strategy* were known to the opponent, then he would in turn choose scissors, stone, and paper with relative frequencies $1 - p_1 - p_2$, p_1, and p_2. Such a conclusion cannot be based on a single instance of the game. Let us then first imagine a row player who—as depicted in Figure 1.2—in the manner of one-sided, shortsighted play that extends over a large number of rounds responds to each action of his opponent in one round by responding with his best reply to that action in the following round.

The same reasoning again leads to a consistent choice with probabilities p_2, $1 - p_1 - p_2$, and p_1 for the actions of the column player. However, this conclusion agrees with our original assumption about the behavior of this player, that is, that the relations $p_1 = p_2$ and $p_2 = 1 - p_1 - p_2$ hold. By doing the arithmetic one obtains for both players the valid mixed strategy $p^\star = \left(\frac{1}{3}, \frac{1}{3}, \frac{1}{3}\right)$, which leads to each possible move being made with equal probability.

What properties can we attribute to the strategy $p^\star$? No other strategy does better against it! Of course, there are other strategies— for example, among (infinitely many) others the so-called *pure strategies*, that is, a strategy based on a fixed principle, such as always showing scissors—that work just as well. If in our case we were not dealing with a zero-sum game, then admittedly we would have some problems in basing the method of play $p^\star$ solely on a randomized point of view and thereby without reference to the shortsighted method of play that we imagined earlier.[4]

At this point, however, the second property of this mixed strategy arrives at just the right time. A player who employs strategy $p^\star$ achieves at least the *security level* of the game. This value is the greatest possible profit that a player can be guaranteed of achieving irrespective of the action of the opponent.

---

[4]The masters of game theory have managed in the most marvelous ways to argue away most of these problems with hair-splitting interpretations. We will have more to say about this later.

It should be noted that none of the pure strategies played in $p^\star$ with positive probability can, taken alone, guarantee the same security level. Therefore, among the best replies to $p^\star$ the only reasonable strategy is the mixed strategy itself.

We shall be devoting considerably more attention to the fundamental impulse toward strategic behavior that we have been investigating in the context of the scissors–stone–paper game. However, let us now take the plunge and establish it in a rather charlatanical manner:[5]

(Tabular) Tips for Glass Bead Players

1. *See through your opponent and be on your guard.*
2. *Never fail to give your best reply.*
3. *Play the game through in your mind.*

1.2. A Game of Hackenbush

Games that proceed by a series of moves require a representation different from that of scissors–stone–paper. In the following game of hackenbush,[6] Ms. Segment and Mr. Point are at work alternately removing, each according to her or his name, an edge (composed of segments or, respectively, points) from a given graph that is anchored to the ground. With each removed edge all other parts are removed that are no longer anchored. The game is over when a player—who is then declared the loser—no longer has a valid move.

We shall now introduce a notation for hackenbush that will allow us to describe the reachable positions of each player during the course of the game. We use a three-part encoding: The first part will always contain a letter that stands for the player whose turn it is, thus $\mathcal{S}$ for Ms. Segment and $\mathcal{P}$ for Mr. Point. The second position indicates which branch of the graph, counting from left to right, has been chosen

---

[5] Credit for the marketing of game theory as a vade mecum for managers must be granted to others [**66**]. One should be particularly curious about upcoming personality changes: Captains of industry who up to now have risked their necks in the art of war of Sun-tzu [**24**] or *Gorin-no-sho* (Miyamoto Musashi's *Book of the Five Rings*) perhaps now will lighten up a bit.

[6] The game of hackenbush is described in [**6, 7**], the bible of combinatorial game theory.

by the player in order to remove the edge that is indicated by the contents of the third position, counting from bottom to top. In the case where a player has no valid move left, each of the last two places will be filled with the digit 0.

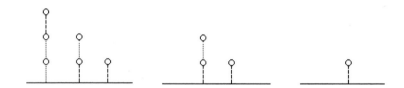

Figure 1.3. A game of hackenbush

A possible sequence of moves in our game (with initial graph given on the left in Figure 1.3) could look like the following: $\mathcal{P}11$, $\mathcal{S}21$, and finally $\mathcal{P}00$. This game is summarized in Figure 1.3. Thanks to his disastrous first move, Mr. Point has gone down to an ignominious defeat. To determine how and whether such a tragic outcome (for Mr. Point) could have been avoided, we shall make use of the *game tree representation* for hackenbush.

The network in Figure 1.4 moves out from a single black node, the *root*. This node is assigned to Mr. Point, who in our game of hackenbush has the first move. All the variant opening moves available to Mr. Point are indicated by directed edges of the game tree that lead out from the root. Each of these edges can be considered the starting point of a particular development of the game.

Since in a game of hackenbush the players alternate moves, it becomes Ms. Segment's move at the goal node at the end of the initial directed edge. Corresponding to the choices of moves available to her, we append new edges to these nodes. When a particular development reaches its conclusion, the winner of the game is recorded at the corresponding terminal node. If Mr. Point is the winner, we indicate this with a utility value of $+1$. Otherwise, Ms. Segment is the winner, and -1 is placed at the terminal node.

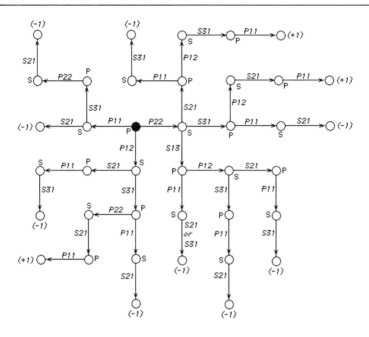

Figure 1.4. The game tree for hackenbush

Hackenbush is an alternate-move, zero-sum game that, in contrast to scissors–stone–paper, has no draw. The basic logical structure of its game tree representation can be transposed to much more complicated game situations. We have the following rules:

Game Tree Representation

1. *A directed edge in the game tree corresponds to one move and connects the "start node" of that edge to a goal node. A sequence of directed edges describes a **path** if for each edge in the sequence (except for the first) the start node is simultaneously the goal node of that edge's predecessor.*

2. *Every path in the game tree corresponds to a **prehistory** or **partial history** of the game.*

3. *A start node that has no prehistory*[7] *is called a* **root**. *If for a particular goal node there is no further history, then such a node is called a* **terminal node**.

4. *If a path begins at the root and ends in a terminal node of the game, then the associated history is called* **terminal**. *Every non-terminal history is associated to a player whose turn it is directly following the goal node of the last edge in the path. Every terminal history determines the corresponding course of the game and thereby determines the player's payoff value.*

5. *If chance is to be integrated into the game, then one simply designates an additional player (Mother Nature, say) whose actions determine the further course of play with fixed probabilities.*

6. *At every node representing the terminus of a particular prehistory the player whose turn it is may freely choose his move from among those available to him.*

The *extensive form*[8] just defined, based on the game tree representation, assumes implicitly that a player knows the entire history that precedes his present action. One therefore speaks also of an extensive game with *perfect* information.

But let us return to Mr. Point and his prospects for victory. In order to assess these correctly we shall follow the various courses the game might take, working our way backward from the terminal nodes. In Figure 1.5 we have first of all labeled all the nodes (clockwise from inside to outside, and at the end we had to make use of capital letters) and then tracked down those nodes of the game tree in which Mr. Point and Ms. Segment have a choice of moves for the last time in the game.

These are the nodes **e**, **f**, **h** for Mr. Point and **d** and **p** for Ms. Segment. Each of these nodes can now be interpreted as the root of a subgame—which is much easier to solve. Thus Mr. Point clearly avoids at all three nodes the move $\mathcal{P}11$, which leads to certain defeat.

---

[7]For the mathematicians among game theorists this is the *empty* prehistory.

[8]The *extensive form* of a game consists of a detailed presentation of the method of play, in particular, the order in which the various players make their moves, as well as information that each player has available at the time of his move. The manner in which the phenomenon of chance enters into play, as well as the stipulation of the payoff values that accrue to the players in the course of the game, complete this description.

For this reason we have deleted the corresponding moves from the game tree in Figure 1.5. We have thereby made (in peculiar agreement with the original rules of hackenbush) entire sequences of moves invalid.

Ms. Segment, on the other hand, is indifferent as to how she moves. Whatever she does at her two nodes will bring her certain victory. For this reason we replace the corresponding subtree by the metanodes **N** and **O**, which are now considered newly created terminal nodes of the game (see Figure 1.6).

After a successful reduction, the backward analysis is continued for the current last choices at nodes **b**, **c** (Ms. Segment) and **g** (Mr. Point). Ms. Segment can coordinate her action to the future behavior

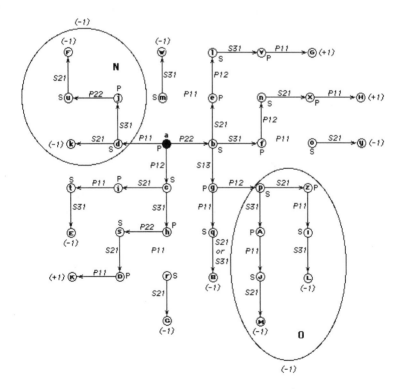

Figure 1.5. Final alternatives in the game tree (actions that do not represent best replies have been deleted)

of her opponent and accordingly hinders the attainment of a subgame position that Mr. Point would prefer. Mr. Point for his part realizes that his actions lead in any case to the same goal, and he remains indifferent.

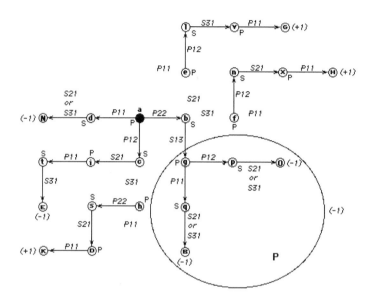

Figure 1.6.   Reduced game tree for hackenbush

The result of these considerations proves, in fact, to be highly unsatisfactory for Mr. Point. If his opponent makes no errors in play, this game of hackenbush—as can be seen in Figure 1.7—is for him a lost cause from the very first move.

It is nonetheless worth noting that a plan of action must contain recommendations even for such nodes that could not be reached at all if the player actually employed his plan. Such unusual plans[9] are to be regarded as strategies in the *extensive sense*.

We can collect into the following rules the additional strategic insights that we have gained from a consideration of the hackenbush example.

---

[9]Even stranger than the plans themselves are many of the current interpretations. More on that in Chapter 7.

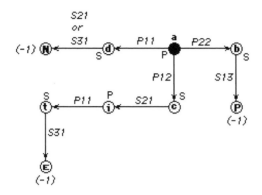

Figure 1.7. Mr. Point's fiasco

(Expanded) Tips for Glass Bead Players

1. *Always look back to plan ahead.*[10]
2. *Consider the impossible; it will exclude itself of its own accord.*[11]

Yet hackenbush players will admittedly scarcely be tempted to make use of a complex game tree representation in evaluating their chances of victory or defeat. The mathematics of combinatorial games has developed special approaches for this purpose (see, for example, [6]) that are particularly well adapted to the peculiarities of hackenbush.

On the other hand, game theory has a completely different view of things. It views as of primary importance the interactive patterns of behavior of the players as they search for a solution (and only secondarily the solution itself). The advantages of this point of view are obvious. The methods of game theory do not depend on the specific conflict situation currently under investigation.

The general validity of its conclusions and concepts must, however, always be subject to the criticism of empirical observation and

---

[10]Whereby we have indeed reached a state of Kierkegaardian proportions: "Life can only be understood backwards, but it must be lived forwards."

[11]Even the master of strategic moves Sherlock Holmes thought only in the other (logical) direction [17]: "When you have eliminated the impossible, whatever remains, however improbable, must be the truth."

experiment and should prove itself not least in the crucible of many specialized applications.

Robert Aumann, one of the influential thinkers on game theory, suggests, on the basis of the above-mentioned arguments, changing the name of the discipline to *interactive decision theory* in order also to clear out the accretions of the idea of "game" (based on the image of parlor games), which in the specialist literature plays a very small role. This opinion seems to be in strong opposition to a well-known statement of Martin Shubik: "I don't believe any game that can't be played as a parlor game."[12]

Between these two positions we shall continue to navigate our lighthearted voyage through the intellectual landscape of the strategic calculus.

1.3. The Paladins of Game Theory

> Let others sing of knights and paladins
> In aged accents and untimely words.
> —**Samuel Daniel,** *Sonnets to Delia*

The mathematical facets of the striving young discipline were honed on the inexhaustible variants of tactical problems. Thus we read in the masterful monographs of such as Melvin Dresher [**23**], Rufus Isaacs [**44**], and Samuel Karlin [**47**] about duels, pursuit games,[13] battles of attrition, and dogfights reminiscent of the Red Baron.

At the beginning of the 1950s it would have sufficed quite well to lay one's hands on a directory of the employees and consultants of the RAND corporation[14] to obtain a who's who of game theory. The zeitgeist was caught in a zero-sum pattern of thought. It was at RAND that Dr. Strangelove and his colleagues learned to stop worrying and love the bomb. It is no wonder, then, that these knights

---

[12] In [**88**] one may marvel at further Shubikian pearls from the treasure-house of game theory.

[13] Usually bearing such cryptic names as "the lady and the lake," "the princess and the monster," or even "the suicidal pedestrian."

[14] The acronym RAND stands for "research and development." In the period directly following the end of World War II RAND was established by the United States Air Force to deal with particular questions of national security. Among those associated with RAND are to be found (among others) the famous names von Neumann, Nash, Kahn, Karlin, Flood, Dresher, Isaacs, Schelling, and Tucker. RAND can also be given credit for significant contributions to the first computers and the internet.

of the holy grail of operations research[15] were capable of thinking the unthinkable:

The RAND Hymn

The RAND Corporation's the boon of the world
They think all day long for a fee.
They sit and play games
About going up in flames
For counters they use you and me....[16]

One can find out more about this RAND episode of game theory in [**73**], an amusing treasury of stories, biographies, and anecdotes from the stone age of the glass bead game, for which we have to thank the pointer to the RAND Hymn. By the end of the decade the initial enthusiasm that had been showered on zero-sum theory was replaced even in military circles by a reserved sobriety. At this time, however, a decisive change had already taken place. With the twenty-seven pages of his doctoral dissertation a young student of mathematics changed the direction in which game theory was to develop. His name was John Forbes Nash.

John Forbes Nash

Born 13 June 1928 in Bluefield, Virginia, Nash studied at Princeton under Albert W. Tuckers. He returned to Princeton as a professor by way of MIT (Massachusetts Institute of Technology). Fate granted him only a brief span for the compelling proof of his multifaceted mathematical talent. In 1959 a serious illness left him incapacitated, and his forays into the world of scientific research became more and more infrequent. In the mid-1980s Nash conquered his affliction and was able to break out of his isolation.

It was Nash who drew the fundamental boundary between co-operative and noncooperative games. We have him to thank for the Nash bargaining solution [**67**], the idea of the Nash program with the

---

[15]The simplex method of linear optimization and the theory of dynamic programming belong to the first developments at RAND. Recently, the field of operations research has been forced to rely on lonely academic outposts.

[16]Text and music by Malvina Reynolds. Copyright 1961 by Schroder Music Co. (ASCAP). Renewed 1989 by Nancy Schimmel.

goal of modeling cooperative situations by means of rules of a newly defined noncooperative game [**69**], and, finally, the plan for a universal solution schema for noncooperative games: *strategic equilibrium* (also called *Nash equilibrium*).[17]

It is impossible to explain the influence of these contributions to modern mathematical economics without pointing to the accomplishments of those who built the proud tower of noncooperative game theory upon Nash's foundations [**68, 69**].

Every game in extensive form possesses a unique normal form representation. Reinhard Selten observed that equilibria of normal form cannot automatically be regarded as reasonable solutions in a corresponding *extensive form*. In his works [**84, 85**] he suggested the first refinements of the concept of equilibrium.

Reinhard Selten

Born 10 October 1930 in Wrocław, Selten felt himself drawn to mathematics at a young age. He spent his student years in Frankfurt, where he wrote his master's thesis under Ewald Burger on a topic in cooperative game theory. After his first publications in the field of experimental economics, Selten established himself within a short time as one of the most innovative researchers in game theory. After holding academic posts in Berlin and Bielefeld, he became professor of economics, in particular economic theory, at the University of Bonn. Particularly noteworthy is Selten's interest in intellectual cooperation, with significant interdisciplinary work in the fields of political science, biology, and psychology.

John Harsanyi extended the validity of equilibrium analysis to the class of games with incomplete information.[18] He showed in [**36**], [**37**], and [**38**] how it is possible using a Bayesian approach to convert an information deficit into a quantifiable uncertainty.[19] In recognition of their groundbreaking work [**68, 69, 84, 85, 36, 37, 38**], the

---

[17]In a Nash equilibrium no player feels the need to alter his position, that is, to play a strategy other than his (so-called) equilibrium strategy, since he can assume that the other players with their strategies will all remain in equilibrium.

[18]In such models certain characteristics of a game, such as the utility values, are not known equally by all players.

[19]We shall see an example of this process in Chapter 8.

Royal Swedish Academy of Sciences awarded the 1994 Nobel Prize in economics to Nash, Harsanyi, and Selten.

John Harsanyi

John Harsanyi was born on 29 May 1920 in Budapest. In the chaotic postwar period he became a political refugee. He studied philosophy in his new home, Australia, and then economics at Stanford University. He then accepted a position at Wayne State University, in Detroit, followed in 1964 by a professorship at the University of California, Berkeley. Among the most important areas in which he did research are bargaining theory and utilitarian ethics.

Chapter 2

Equilibrium and Game as Metaphor

> In battle or business, whatever the game,
> In law or in love, it is ever the same;
> In the struggle for power, or the scramble for pelf,
> Let this be your motto — Rely on yourself!
> —**John Godfrey Saxe,** *The Game of Life*

Among the seventy-eight different non-zero-sum, two-person, normal-form games where both players have only two pure strategies from which to choose, one can discover some of the most fascinating social conflict situations generally associated with the notion of game theory.

In this chapter we shall interpret three famous social conflict situations and their Nash equilibria. All three share the property of symmetry, which we have already encountered with respect to the scissors–stone–paper game. The essential difference from that game of hand signals is that here we are dealing with a non-zero-sum situation: The players can both emerge winners, or they can both lose.

Each player has two options: The first option, which we shall call the *lion strategy*, calls for aggressive, noncooperative behavior, while the second, the *lamb strategy*, represents a fundamentally cooperative attitude.

In Figure 2.1 we have indicated the payoff values that accrue to the first player (the row player) for the various strategic constellations. On the basis of symmetry we may determine the payoff values for the second player as we did in Section 1.1.

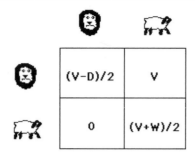

(V−D)/2	V
0	(V+W)/2

V · · ·   a winner's increase in fitness

D · · ·   a loser's decrease in fitness

W/2 · · ·   average increase in fitness due to being a peacemaker

Figure 2.1.   The lion–lamb game

If the column player assumes a leonine attitude, while the row player is as meek as a lamb, then his payoff value corresponds to that of a row player playing the lion who encounters a column player taking the role of a lamb, and vice versa.

The lion–lamb game can serve—not surprisingly, considering the name—as a metaphor for Darwin's theory of evolution. In a host of books on game theory this appears—metamorphosed into the form of other animals—as the hawk–dove game. There are many possible reasons for that name, but they are certainly not of an ornithological nature.[1] Doves, primarily because of their highly aggressive nature—thus Richard Dawkins in *The Selfish Gene* [**21**]—are hardly to be put forward as a credible model of cooperative behavior.

The evolutionary parable is quickly told. Two randomly selected individuals in a population become embroiled in conflict over some prize. Sole possession of this resource will increase the individual's Darwinian fitness[2] by an amount V.

---

[1] "Ornithologists know all about the bird, but their nomenclature is absurd" (Ogden Nash).

[2] By this we shall understand the expected number of offspring of an individual in the given population.

While a "lamb" surrenders possession of the prize to a "lion" without a struggle, two "lions" will always fight it out to the finish. If we let D denote the amount by which a defeated individual's **Darwinian** fitness is reduced in the course of such a struggle, then **after** an encounter with another "lion," a "lion" can reckon on an average change of fitness of $(V - D)/2$. Two "lambs" will simply divide the resource equally without a fight. For this reason we can expect **an** average increase in fitness for two "lambs" of $(V + W)/2$, where $W/2$ denotes the average increase in fitness due to being a peacemaker.

The games that we shall now investigate can be understood **at** once in terms of the game table of the lion–lamb game with the assignment of specific values of the parameters.

Specific Cases of the Lion–Lamb Game

1. $V = 4$, $D = 2$, $W = 0$: *dilemma of the arms race;*
2. $V = 2$, $D = 4$, $W = 0$: *game of chicken;*
3. $V = 4$, $D = 0$, $W = 6$: *(Rousseau's) deer hunt parable.*

We shall employ, as is usual in game theory, the bimatrix **representation**. In the lower left-hand corner of each cell we write **the** payoff to the row player, while that of the column player appears **in** the upper right-hand corner. The given payoff values do not represent average changes in fitness. They are merely an expression of the ordering[3] of the possible outcomes of the game according to the valuation of the particular players.

2.1. The Arms Race Dilemma

In England baut man flugs zwei Dreadnoughts mehr.
Im Oberhause stürmen die Debatten.
Es hetzt die Presse gegen Deutschlands Heer.
Erregt kauft ganz Europa Panzerplatten.[4]
  —**Ludwig Rubiner et al.,** *Auf Helgoland*

---

[3]In game theory so-called cardinal utility values are used. They express the amount by which a particular outcome of the game is preferred by a player. Since neither the origin of the utility scale nor the values of the individual utilities really matter, an affine (i.e., positive linear) transformation of the utility values does not produce a fundamental change in the nature of the game.

[4]In England they have built two dreadnoughts more./ The House of Lords is stormily debating./ The press screams, "Germany is plotting war!"/ While all of Europe buys up armor-plating. [Translation by David Kramer.]

Two of the smallest great powers, Lilliput and Blefuscu, have become embroiled in a dispute over the important ideological question of which end of a soft-boiled egg—the broad or the narrow—should be sliced off in order for a diner to achieve optimal enjoyment of his breakfast. Mutual disarmament by these two political entities could actually have the effect of exacerbating the conflict. However, influential groups within each country believe that a one-sided disarmament by the opponent with a concomitant increase in armaments for their own country represents the best of all possible worlds. In Figure 2.2 we present each strategic constellation—from the point of view of each competitor—according to its degree of stability.

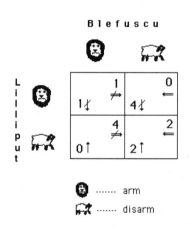

Figure 2.2.   The arms race as a bimatrix game

If a country can expect an improvement in its strategic position, then we indicate this in the appropriate cell of the bimatrix by an arrow[5] that points in the direction of a higher payoff. If, on the other hand, a worsening can be expected, then we mark the corresponding cell with a struck-through arrow. Column players always move horizontally, while row players move vertically within the bimatrix.

There is a unique ending to the game in which neither nation feels the need to play a strategy other than the one that it has chosen, insofar as it must assume that its adversary will continue its strategy.

---

[5]A double arrow for Blefuscu, a single arrow for Lilliput.

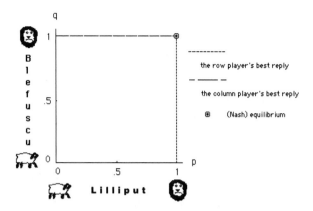

Figure 2.3. Best replies in the arms race dilemma

This ending corresponds to the Nash equilibrium (lion, lion), whereby both sides arm themselves.

We can also argue as follows why an arms race is the only possible solution to the conflict between Lilliput and Blefuscu. No player is prepared to disarm, since—whatever the opponent's action—the associated outcomes to the game yield a worse payoff than what can be accomplished by arming. The option of arming is then for both players under all circumstances the unique best reply. In Figure 2.3 we have recorded this situation graphically.[6]

Another way of saying this is that from the point of view of each player the strategy of arming (lion) strongly dominates the strategy of disarming (lamb).[7] The case (lion, lion) will be called an equilibrium[8] in strongly dominant strategies.

The only strategies that can never be part of a strategic equilibrium[9] are the strongly dominated ones. Players who use them

---

[6]With probabilities p and q the row and column players choose to arm.

[7]If whatever the opponent's action the respective game outcomes of a strategy s have at most as great a payoff as those of another strategy t, but s is not strongly dominated by t, then one speaks of a weak dominance holding between s and t.

[8]Each such equilibrium is also always a Nash equilibrium.

[9]Strategic equilibria can nevertheless certainly contain weakly dominated strategies.

are acting irrationally. We may add to our list of tips an additional prohibition:

Extended (Tabular) Tips for Glass Bead Players

1. *See through your opponent and be on your guard.*
2. *Never fail to give your best reply.*
   (a) *Never use a strongly dominated strategy, since it can never be a component of a best reply.*
3. *Play the game through in your mind.*

Since every rational player heeds tip 2(a), by eliminating strongly dominated strategies one obtains a simplified yet equivalent—in terms of solutions—representation of the original game. The process of deletion is initiated by an arbitrary player, and then the players alternate deleting strategies until no further reduction of the simplified game can be obtained.

In games with more than two options a pure strategy can certainly be dominated by a mixed strategy. This must also be taken into consideration in the deletion process.

In spite of the uniqueness of the Nash equilibrium, the world of our bimatrix game does not seem quite in order. The cause is the cooperative game outcome (lamb, lamb),[10] which is valued by both players more highly than the equilibrium strategy (lion, lion). Cooperation cannot, however, be achieved if the players—in flagrant violation of the assumption of rationality—stray from the path of the best reply.

This paradox indicates, among other things, that with the arms race dilemma we have investigated only one of the multitudinous disguises of the prisoner's dilemma (see [**75**]). In Part II we shall relate this perhaps most famous myth of game theory, which for a long—and for game theory fruitful—time drove early investigators to distraction.

Finally, the problem of rationality and the paradoxes that it creates is addressed in the chapter on game-theoretic scholasticism.

---

[10]This outcome is, in fact, *Pareto efficient*, that is, each alternative outcome that is valued more highly by at least one player will, on the other hand, be valued less by at least one opponent.

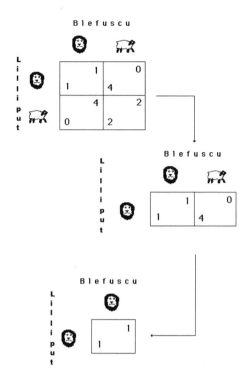

Figure 2.4. Repeated deletion of strongly dominated strategies

2.2. Rebel Without a Cause

> When everyone is courageous, that is sufficient reason to worry.
> —**Gabriel Laub,** *Thinking Ruins Character*

> Where would the heroes be without cowards?
> —**Werner Mitsch,** *Pro- and Contradictions*

In the film *Rebel Without a Cause* James Dean plays a teenager spoiled by prosperity who is challenged to a deadly car race. The two contestants race in stolen cars toward a precipice. The loser (declared a *chicken*) is the first to jump out of his moving car. In the film Dean is beaten by his opponent, the sleeve of whose leather jacket gets caught in the door handle, plunging the wearer into the abyss.

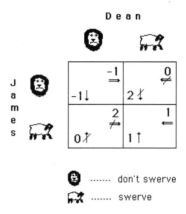

Dean

⚫ ······· don't swerve

🐑 ······· swerve

Figure 2.5. The game of chicken

A suitable translation of these conditions would lead to a so-called timing game.[11] The philosopher Bertrand Russell used a much simpler variant as a metaphor for the nuclear equilibrium of nuclear terror.

Bertrand Russell's Chicken Variant

On a country road closed to general traffic two cars are racing toward each other. Both drivers are racing their cars down the middle of the road on a collision course. The loser—the chicken—is the first to turn aside.

In Figure 2.5 we have formulated this as a bimatrix game. The valuation of the game outcomes yields the following (symmetric) scale:

1. Surviving hero

2. Just as much a chicken as the other

3. Coward (but alive)

4. Dead hero

---

[11] These might perhaps, in the sense of nineteenth-century romanticism, be called duels. In Section 4.1 we shall involve ourselves (with a conspicuous absence of mathematical seconds) in such affairs of honor.

A stability analysis using our tried and true arrow diagrams shows that there are two possible outcomes for which no player feels the necessity to change his current pure strategy unilaterally. But are these really the only strategic equilibria of our bimatrix? Wherever dominance criteria do not allow a simplification of the game, this question can be answered by introducing reaction functions.[12] On account of the symmetry of the game of chicken it suffices to consider the best reply of the row player.

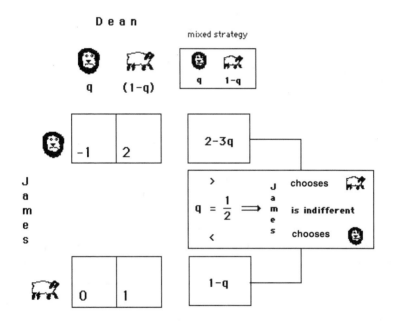

Figure 2.6. The row player's best reply

Now, let q be the probability that the column player chooses the role of the hero ("lion" strategy). What options does the row player have? If he chooses the pure strategy "lion," then his expected utility is as follows:

$$(-1) \times q + 2 \times (1 - q) = 2 - 3q.$$

---

[12]In the future we shall employ the concept of reaction as a synonym for the best reply.

Otherwise, his pure choice of the role of coward leads to a utility value
of $1 - q$.

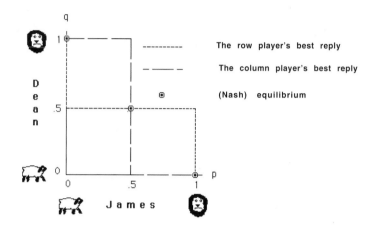

Figure 2.7. Best replies in the game of chicken

A rational player will always respond according to his best reply
and thus always reply with the strategy with the greatest utility. This
is "lamb" for $q > \frac{1}{2}$, while for $q < \frac{1}{2}$ it is "lion." Only in the case
$q = \frac{1}{2}$ is the uniqueness of the best reply lost. The row player is
indifferent as to which of the two pure strategies he chooses. Indeed,
with every mixed strategy he achieves exactly the same utility value.

In Figure 2.6 we have shown the considerations that lead to the
best reply of the row player. The column player reacts analogously in
accordance with the probability p with which the row player assumes
the hero's mantle.

A comparison of best answers in the game of chicken appears in
Figure 2.7. The intersections of the two reaction curves represent the
complete set of Nash equilibria[13] of the game. In addition to the already discussed equilibria in pure strategies, a symmetric equilibrium
can be demonstrated in the case of complete mixed strategies

---

[13]Every bimatrix game (of finite dimension) possesses at least one Nash equilibrium (in mixed strategies). The number of equilibria in a bimatrix game is (up to
nongeneric exceptions) always odd.

Can one say that the players in a simultaneous game of chicken know not what they do? In the case of asymmetric equilibria the hero has to rely on the fact that his opponent is taking the role of the coward. But how can he be sure of this? A symmetric equilibrium, on the other hand, presupposes that each player chooses his strategy—without his opponent knowing what it is—from the flip of a coin. This may indeed be *cool*, but it is hardly efficient.

No wonder, then, that the moderate adepts of game theory vote for a shift of possible coordination efforts to the (nonexistent) pregame. In his description of chicken as a metaphor for the confrontations of the nuclear age [**45, 46**], Herman Kahn has created a truly devilish metaphor.

Metaphor for the Game of Chicken

Right now we are in the last phase of the race. James Dean can already make out the features of his opponent as he races toward him at breakneck speed. Only three, perhaps two, seconds until the fatal crash. Suddenly, Dean rolls down his window and ostentatiously throws the steering wheel out of the car.

A hero is born! The gesture is completely convincing. James no longer has the option of turning aside. But what would happen if his counterpart—a Dean clone—following the same basic strategy and at the same instant, throws his steering wheel out the window?

Underhanded tricks, threatening gestures, idle chatter,[14] can be taken seriously in the pregame phase only if they are defined as valid rules of an extended game. In principle, Robert Aumann's metaphysically inclined concept of a *correlated equilibrium* [**1**] also rests on such an extension.

In such a case the players would have abandoned all control over the game. Thus in the game of *ornitheios*—if we may for the moment employ the ancient Greek word for "chicken"—the misguided teenagers Harmodius and Aristogeiton might, just before climbing into their chariots, consult the Delphic Oracle. Both know that the oracle has probably already chosen one of the three marked outcomes

---

[14]Known as "cheap talk" in the game-theoretic literature.

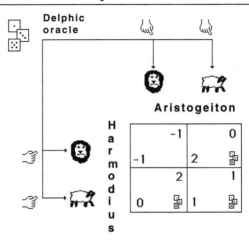

Figure 2.8. How a correlated equilibrium can be reached

(as depicted in Figure 2.8 with pictures of dice). Yet the oracle, as usual, remains quite vague and secretly divulges to each player the pure strategy that he is obliged to adopt.

Will Harmodius and Aristogeiton heed the words of the oracle?[15] If they do, then surely not out of fear of the gods, or because it is written in their horoscopes. The best reason of all is a game-theoretic one: They do it because to do so is a rational course of action.

If the "lion" strategy was recommended to Harmodius, then he knows for certain that his opponent will assume the role of the coward. He will therefore heed the oracle. But if the oracle murmurs "lamb," then Harmodius will decide either on lion or lamb, each with probability $\frac{1}{2}$. His expected utility value of $\left(0 \times \frac{1}{2}\right) + \left(1 \times \frac{1}{2}\right) = \frac{1}{2}$ cannot be improved by a change of strategy. Thus these instructions will also be obeyed. As the most pleasant result of this obedience, each player can look forward to an expected utility of $\left(2 \times \frac{1}{3}\right) + \left(\frac{1}{2} \times \frac{2}{3}\right) = 1$; twice as much as in the symmetric Nash equilibrium.

None of this, however, alters the fact that in the original formulation of the game only the three Nash equilibria come into question

---

[15]In chicken-like situations, for example between union and management, the coordinating role of the oracle can be assumed by a mediator or arbitrator.

as solutions. Evolutionary arguments, as we shall see in Section 2.4, nonetheless reject the asymmetric equilibria and suggest looking on symmetric equilibria in mixed strategies as the solution to the game.

In day-to-day politics asymmetric equilibria have passed the test with flying colors. At the high point of the Cuban missile crisis, Khrushchev, roaring like a young hooligan in a big, black, government-issue ZiL toward Kennedy, turned aside a few seconds before the big bang. In this connection we must all certainly be thankful that the principle of repetition, or that of the Nash equilibrium in mixed strategies, did not come into play.

2.3. The Parable of the Deer Hunt

> Es gingen drei Jäger wohl auf die Birsch,
> Sie wollten erjagen den weissen Hirsch,
> Sie wollten erjagen den weissen Hirsch.
> Husch husch! Piff paff! Trara!
> —**Ludwig Uhland,** "Der weisse Hirsch"

> CURIO. Will you go hunt, my lord?
> DUKE. What, Curio?
> CURIO. The hart.
> DUKE. Why, so I do, the noblest that I have.
> —**William Shakespeare,** *Twelfth Night* I.i.

Jean Jacques Rousseau, one of the most influential representatives of the political philosophy of the eighteenth century, came up with the following analogy of the conflict between individual goals and the common will in his discourse on the origins of inequality:

Rousseau's Parable of the Deer Hunt

A group of hunters has succeeded in surrounding a deer as well as several rabbits. Cornered, the animals attempt to break free. Each hunter now has a choice: to let the rabbits go and together prevent the deer's escape or to settle for second best and go after a rabbit, letting the greater prize leap from their grasp. The deer can be captured only if each hunter resists the temptation to go for the easier quarry. If even a single hunter has too great a hankering after hasenpfeffer, then all who promote the general welfare will definitely get the worst of it.

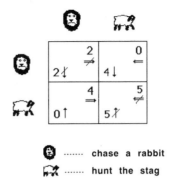

Figure 2.9. The bimatrix of the deer hunt game

The analogy of the deer hunt applies to many social predicaments. Poundstone [**73**] makes the dramatic analogy with the mutiny on the *Bounty*. If a certain critical number of the ship's crew refuse to take part in the revolt against Captain Bligh, then the mutineers led by Fletcher Christian, the first mate, will fail.

A modern example might be that of a group of congressmen who for understandable reasons come to an agreement to oppose a bill that would result in a reduction in their salary. But any individual member of Congress can only with great difficulty resist the temptation to support the bill in the interest of his own reputation—to the extent that he estimates the danger that the bill will pass as being not too great. The greater the number of representatives who give in to this temptation, the more likely it is that the unpleasant decision will be taken (by a simple majority).

In Figure 2.9 we have reduced the deer hunt parable to the level of a two-person game, the so-called deer hunt game.

As pure Nash equilibria the symmetric pairs (lion, lion) and (lamb, lamb) commend themselves.

A graphical analysis of the best replies in Figure 2.10 confirms this recommendation and identifies an additional symmetric equilibrium in mixed strategies. Moreover, (lamb, lamb) dominates (from the

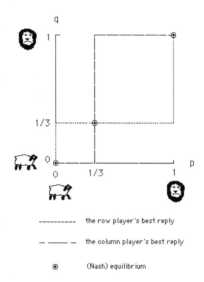

Figure 2.10.   Best replies for the stag-hunt game

point of view of both players)[16] over both other equilibria. Is this equilibrium finally the solution to the game? Harsanyi and Selten would, in the sense of their complicated theory of equilibrium selection [**39**],[17] suggest that the game plan (lamb, lamb) is *risk dominated*[18] by (lion, lion) and thereby has more to lose by an error in coordination on the part of the opponent.

2.4. The Revenge of the Mutants

> Evolution is not a force but a process; not a cause but a law.
> —**John, Viscount Morley of Blackburn,** *On Compromise*

---

[16]In this case one speaks of *Pareto dominance*.

[17]Truly a game-theoretic witch's hammer (*malleus maleficarum*) that contains all inquisitorial investigations that (doubtless?) should lead to the unique solution of the game.

[18]In our simple game the risk dominance can be explained as follows. Someone playing the lion will regret his choice of strategy only if his opponent chooses the lamb strategy with a probability greater then $\frac{2}{3}$. This value of $\frac{2}{3}$ gives the oppositional power of (lion, lion) to the game plan (lamb, lamb). On the other hand, the oppositional power of (lamb, lamb) to the game plan (lion, lion), is, at $\frac{1}{3}$, considerably less.

We now would like to interpret the games of the foregoing sections in the context of evolutionary conflict. For chicken, for example, we obtain the fitness matrix depicted in Figure 2.11.

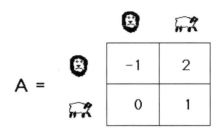

Figure 2.11. Chicken as evolutionary game

Here the numbers denote the average values by which the Darwinian fitness of an individual or population rises or falls when it exhibits a particular behavior (lion or lamb) in the course of a confrontation in which it faces an opponent who for his own behavior possesses only these two options.

Evolutionary conflicts actually only appear to be between individuals of a population. The true opponents in this game are the puppeteers and not the marionettes. Without being aware of it,[19] the individuals are governed by heritable behavioral (in the case of *Homo sapiens* also intellectually transmittable) algorithms.

The puppeteers are also known as *replicators*, since they can reproduce themselves only indirectly, by way of their marionettes. Richard Dawkins's selfish gene and—as new recruit—his meme belong to the pantheon of replicators in [**21**].

Let us introduce the following notation:

A: The fitness matrix of an evolutionary conflict

$\tilde{p}$: A mixed strategy that has established itself as normal behavior (induced by the replicator N).

---

[19]However, there are certain populations whose individuals write books on this theme.

p: A mixed strategy that as a result of a mutation (induced by the replicator F) in a small fraction ϵ of the population appears as aberrant behavior.

An individual whose opponent in a conflict is chosen at random meets a marionette of replicator F with probability ϵ and one of replicator N with probability $1 - \epsilon$. The average fitness then comes to $(1-\epsilon)\tilde{p}^{\mathrm{t}}A\tilde{p}+\epsilon\tilde{p}^{\mathrm{t}}Ap$ for the normal replicator and $(1-\epsilon)p^{\mathrm{t}}A\tilde{p}+\epsilon p^{\mathrm{t}}Ap$ for the mutant.

Maynard Smith [58] has called a strategy *evolutionarily stable* if the replicator that induces it evinces a greater average fitness than every other replicator that appears in a sufficiently small part $\epsilon > 0$ of the population.

ESS (Evolutionarily Stable Strategy)

In an evolutionary conflict determined by the fitness matrix A the mixed strategy $\tilde{p}$ is evolutionarily stable if and only if:

- *It is a best reply to itself.*
- *It does better against every mutant strategy $\hat{p}$ that is the best reply to $\tilde{p}$ than $\hat{p}$ does against itself.*

Evolutionarily stable strategies are to be found only among those strategies that are part of a symmetric Nash equilibrium. In the game of chicken we have a unique candidate: the mixed strategy that decides with equal probability between "lion" and "lamb."

Every behavior proves to be a best reply to this strategy. The fitness value $(3-4p)/2$, which the strategy $\left(\frac{1}{2}, \frac{1}{2}\right)$ achieves against any other strategy $(p, 1-p)$, is nonetheless greater by exactly $2\left(p - \frac{1}{2}\right)^2$ than the fitness value $1 - 2p^2$ that $(p, 1-p)$ achieves against itself.

However, this weighing of strategies creates a false picture of the evolutionary game. The distinguishing mark of evolutionarily stable strategies is of a dynamic nature. We can regard them on the one hand as strategies that are able successfully to resist the infiltration of mutants into the pool of strategies. On the other hand, if we imagine evolutionary development as a dynamical system that always moves in the direction of those pure options that promise a local increase in fitness, then ESS prove to be evolutionarily robust.[20]

As a prototype of successful learning behavior this dynamic viewpoint has also influenced other areas of game theory. The most authoritative source is Fudenberg and Levine [29]. In [19] and [20] Dawid and Mehlmann consider genetically coded populations of mixed strategies whose individual fitness is influenced by the present state of the population. A genetic algorithm plays the role of evolution using selection, cross-breeding, and mutation.

---

[20]The mathematicians would call them—totally prosaically—(locally) asymptotically stable stationary points of the replicator dynamics. For 2×2 fitness matrices the ESS correspond to the asymptotically stable points of the dynamics. But as soon as more than two options come into play, then there can certainly be asymptotically stable points that do not correspond to anything evolutionarily stable. In this case even the existence of ESS is no longer ensured.

Chapter 3

In the Forest of Game Trees

A fool sees not the same tree that a wise man sees.
—**William Blake,** *The Marriage of Heaven and Hell*

The first game trees stretched forth their leafy branches in von Neuman and Morgenstern's monograph [**71**]. Kuhn's [**54**] concept of strategies for these complicated positional games was rather simple: a function specifying the player's action in each of his information sets.

While it is always possible to bring the formulation of a game tree into abstract normal form in order to carry out a successful search for equilibria, mixed equilibria in normal form provide no immediately understandable pattern of behavior in extensive alternate-move games.

It was again Kuhn who showed the way out of this dilemma. In [**54**] he showed that for the class of extensive games in which all players are characterized by perfect memory[1] there exists an equivalent way of representing mixed strategies. For every information set in which it is his move, a *behavioral strategy* specifies the probability for choosing each of the player's available actions (a probability distribution over the player's available actions).

---

[1] As regards their past knowldege and moves they have already made.

In this chapter we shall explain the strategic properties of extensive modes of play with the help of examples of game situations. The credibility of strategic equilibria will be put under the microscope in the first of these games. We have Selten [84, 85] to thank for the insight that many equilibria in unreachable parts of the game tree yield questionable, since nonequilibrium, recommendations.

In what follows we shall turn our attention to the bestiary of game theory. Rosenthal's "centipede," Selten's "horse," as well as Kohlberg's "dalek" illuminate many of the ideas of game theory, such as those of further refinement theorems and backward and forward induction.

3.1. The Strange Case of Lord Strange

> He said, "giue me my battell axe in my hand,
> sett the crowne of England on my head soe hye!
> ffor by him that shope both sea and Land,
> King of England this day I will dye!"
> — *Ballad of Bosworth Field*

He was awakened in the morning twilight from a fitful sleep. Shivering, the last Plantagenet paced before the royal war tent and looked anxiously across at the enemy. The view of the military map as shown in Figure 3.1 was spread out before the battle-tried leaders of the advance guard.

The army of rebels was encamped in disarray to the southwest of the swamp. At a suitably respectful distance from the Tudors' right flank the armies of the Stanleys awaited what was to come. Lost in thought, Richard fingered his nonexistent hump and wrinkled his brow into careworn creases.

Could he, when all was said and done, trust this race of Stanleys, this pillar of his kingdom upon whom honors and benefices had been heaped? William's treachery seemed certain. Even if his banishment had come too late, his three thousand men would hardly jeopardize Richard's situation. The case of Lord Stanley, the constable, was quite different. Whoever could depend on his support would surely win the day.

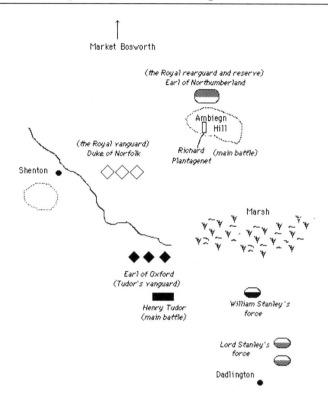

Figure 3.1. The Battle of Bosworth Field (22 August 1485)

Richard played his last trump. Before the morning had passed he sent a messenger to Lord Stanley. The message was clear and unambiguous. Should he hesitate to support his king, then Lord Strange, the king's hostage and Stanley's son, would forfeit his head.

In Figure 3.2 are given, corresponding to the three possible outcomes of the game, the valuations according to each player. Stanley clearly prefers to withhold support if he can assume that Richard will not carry out his threat. For this reason this outcome is given, from Stanley's point of view, the utility value 0. For Richard this outcome with value 0 is only the second-best outcome. He would most like to have Stanley's support; he would value this latter outcome at 5, while

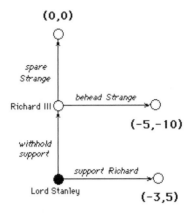

Figure 3.2.   The game for Richard's last trump

Stanley gives it the value −3. Finally, both players value the execution of the hostage as the worst outcome. For Richard the utility is −10, while for Stanley it is −5.

Will the king's threat fall on fertile soil? A brief glance at the normal form representation associated to the game tree in Figure 3.2 lets us imagine the grisly outcome.

Two Nash equilibria[2] are circled in this bimatrix. In the first of these equilibria Lord Stanley gives in to Richard's threat[3] and decides to support him. The second equilibrium describes a Stanley who withholds support and a king who then does not dare to carry out his threat.

How are these two equilibrium solutions to be evaluated? The first of the equilibria is maintained only by an empty threat and therefore should be eliminated from the category of reasonable solutions.[4] A glance back at the game tree in Figure 3.2 allows us to recognize the correct way to proceed: the technique of backward induction.

---

[2]More precisely, the outcomes of the equilibria.

[3]In Figure 3.4 additional Nash equilibria in mixed strategies are described. Richard threatens in these equilibria to behead the hostage with probability $\frac{3}{5} \leq q < 1$ if Stanley does not support him.

[4]Together with the other threat equilibria in Figure 3.4, which are equally unbelievable.

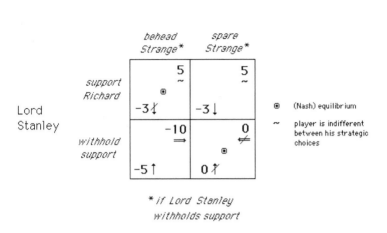

Figure 3.3. Richard's last trump—the corresponding normal form

We first consider the subgame whose root coincides with Richard's sole decision node. Confronted with the choice of whether to carry out his threat, Richard has only one remaining option: to spare Strange. Once the empty threat has been eliminated from the subgame tree on the grounds of its being a strictly dominated action, then Lord Stanley will withhold his support in the root of the original game. The resulting equilibrium (*withhold support, spare Strange*) is the only one that fulfills the property of subgame perfection.[5]

A subgame perfect equilibrium exists in every finite game tree with perfect information. For the case that no player is indifferent with regard to two different outcomes, then even the uniqueness of the subgame perfect equilibrium can be demonstrated. In the associated

---

[5] A subgame perfect equilibrium recommends only such plans of action that form an equilibrium in an arbitrary subgame of the original game (even in those that remain unreached in the corresponding course of the game). We have Selten [**84**] to thank for this fundamental refinement of the Nash equilibrium.

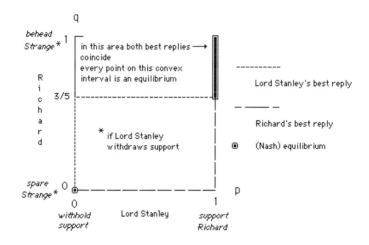

Figure 3.4. Richard's last trump—Nash equilibria

(reduced) normal form such an equilibrium will on no account contain weakly dominated strategies.[6]

Stanley's reply to Richard was short and contemptible: "I have other sons."

We assume that the bearer of this bad news returned with mixed feelings. We would like now for a brief moment to offer a different, game-theoretically motivated, turn to the actual events. On his daring ride across Redmore Plain[7] the messenger, together with his message, was overtaken by Breton marauders. This constructed incident has the most interesting consequences for the "game for Richard's last trump."

Richard has not observed his opponent's first move. His information set now consists of the two decision nodes that are connected by a dashed line in Figure 3.5. Such a game tree describes an extensive game with imperfect information.

---

[6]In Figure 3.3 this would be *Execute Strange if Stanley withholds support.*

[7]It was by this name that the battlefield near Bosworth was originally known.

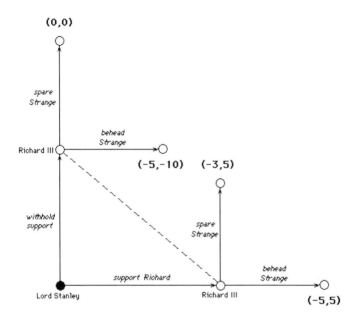

Figure 3.5. Richard's last trump—the messenger's misfortune

All the nodes that belong to the same information set of a player must lead to the same number and type of continuations.[8] The outcomes, however, that arise from the application of identical strategies at different nodes of an information set can be valued completely differently. Thus in Figure 3.5 the beheading of Strange works to Richard's disadvantage only in the case that Stanley withholds support (and Richard knows of this).[9]

In Figure 3.6 the normal form of a game with imperfect information shows two familiar equilibria.[10] The first equilibrium now consists entirely of weakly dominated strategies. We shall scarcely be able to eliminate it by means of backward induction. Namely, the

---

[8]In our example these are the actions *spare Strange* and *behead Strange*.

[9]Only in this case does Stanley have the option of throwing in his lot with the Tudors. If, on the other hand, Stanley has decided to support Richard, then (at least we assume so) a change of sides is out of the question. For this reason Richard would be indifferent as to his options (Figure 3.6) if he could be sure of Stanley's support.

[10]And no others, which indicates a nongeneric case.

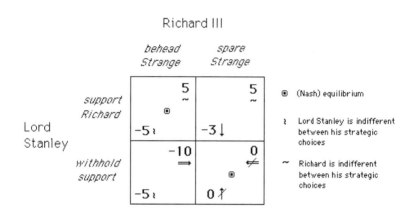

Figure 3.6.   The messenger's misfortune—normal form

game tree in Figure 3.5 possesses no subgame tree other than itself.[11] This means that both equilibria are subgame perfect.

The only way that offers itself out of this dreadful state consists in a further refinement of the characteristic of the equilibrium. From among the possibilities open to us we shall for the time being bring the historically oldest into play.

In [**85**] Selten investigates the question of the robustness of an equilibrium with respect to possible errors that the players can make in choosing their actions. It is not here a question of errors in thought; we are thinking rather of a player who with trembling hand presses the wrong button on the elevator and ends up on the wrong floor.

Every equilibrium that possesses this robustness property[12] must consist of best replies to defective action plans that—if it is possible to subdue the trembling step by step until it entirely disappears—for their part converge to the strategic components of the equilibrium.

---

[11]Note that a decision node can be a root of its own subtree only when the information set of the player whose turn it is contains no other node.

[12]Selten calls this *perfection*, or often, to distinguish it from subgame perfection, *trembling hand perfection*

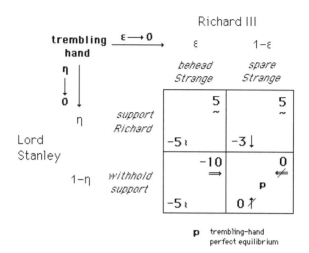

Figure 3.7. A (trembling hand) perfect equilibrium

Even if Richard is now prepared to spare Strange if need be, Stanley, on the other hand, is firmly decided to withhold his support. We know now that each of these two strategies is the best reply to the other. But what happens if one of the opponents develops a tremor?

Stanley's light tremor leads his troops with the low probability η over to Richard's side. The best reply to this completely mixed strategy is nonetheless the same for every $\eta < 1$: *Spare Strange*. On the other hand, if Richard's royal hand trembles, then Stanley's head will roll off his shoulders with the small probability ϵ. But none of this can shake Stanley's resolve. He stands, with probability $\epsilon < 1$, by his determination to withhold assistance.

Moreover, at least one sequence of pairs of fully mixed tremor strategies converges to the pair of these best replies if the trembling completely disappears.[13]

In Figure 3.7 a unique (trembling hand) perfect Nash equilibrium can be identified. What has happened to the other Nash equilibrium

---

[13] Choose, for example, $\eta = 2\epsilon$ to guarantee this and then let ϵ approach 0.

in Figure 3.6? In accordance with the refinement rules that we employed it must be excluded. In normal form games contested by two persons with finitely many choices of action a Nash equilibrium is (trembling hand) perfect if and only if it contains no weakly dominated strategies.

3.2. A Game-Theoretic Bestiary

> Wild, dark times are rumbling toward us,
> and the prophet who wishes to write a new
> apocalypse will have to invent entirely new
> beasts ...
> —**Heinrich Heine,** *Lutezia*

The first of the three game-theoretic beasts that we shall consider in the following seems, based on its appearance, to have its origins in the TV series "Dr. Who."[14] The game tree in Figure 3.8 is modeled more or less on the actual daleks—merciless robots bent on world conquest (in our game, however, apparently interested only in utility values [8]).

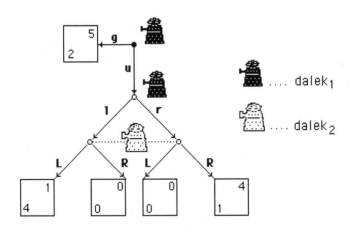

Figure 3.8. Kohlberg's daleks

Dalek 1 determines at the root of the game whether the game ends at once (move **g**), or continues (move **u**). If he (she? it?) moves

---

[14]Whose eponymous hero in *Dr. Who and the Daleks* (1965) and *Daleks—Invasion Earth 2150 A.D.* (1966) was played masterfully by Peter Cushing.

downwards, then he is allowed to make the next move as well and move either left (move **l**) or right (move **r**). Only then does the imperfectly informed dalek 2 have a turn.

The (pure) normal form strategies determine the moves that a dalek should make as the game progresses in its corresponding (ordered temporally) information sets. In Figure 3.9 we have indicated both the complete and reduced normal forms.

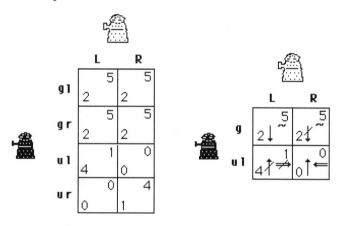

Figure 3.9. Daleks—(complete and reduced) normal form

Although upon taking move **g** dalek 1 has no more turns, the plan of action that we generally associate with a strategy demands that we also record his action in the unreachable second decision node. Since moves **gl** and **gr** lead to the same payoff values (from the point of view of both players) and moreover, **ur** is strictly dominated, the reduced normal form appears to suffice for carrying out an equilibrium analysis.

The following Nash equilibria can be established for pure strategies: (**gl**, **R**), (**gr** , **R**), and (**ul**, **L**). Among these only the last two are (trembling hand) perfect (and thereby subgame perfect).[15]

Nevertheless, we should note that a tremor in the dalek game has its own peculiarities. The errors in judgment that dalek 1 makes in

---

[15]The equilibrium (**gl**, **R**) cannot possibly be subgame perfect, since (**l**, **R**) is not an equilibrium of the subgame begun by dalek 1 in the second decision node.

the two information sets in which it is his turn must be uncorrelated. In order to give this an adequate mathematical formulation, we must let dalek 1 be represented by an agent in each information set associated to him. The normal form that thereby results, called *agent normal form*, will henceforth allow a correct trembling, which for the equilibrium (**gr**, **R**) can be modeled as follows.

The first agent can fail to make the move **g** with the small probability ϵ and play **u** instead. If **u** is played, then the second agent now has a turn and fails to make the move **r** with the small probability δ. Finally, dalek 2 trembles a bit and fails to make the move **R** with the small probability η.

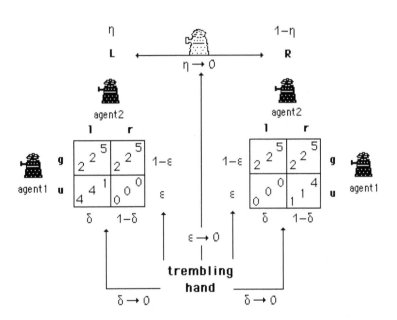

Figure 3.10. A touch of dalek trembling in agent normal form

In Figure 3.10 both agents have at their disposal the utility values[16] of their patron dalek 1. For sufficiently small probabilities of

---

[16] Respectively the values in the lower left corner and in the middle of each matrix cell.

a false move (such as, for example, $\eta < \frac{1}{5}$ and $\delta < \frac{4}{5}$) that then approach zero[17] the strategies **g** and **r** form for both agents, as well as **R** for dalek 2, a (trembling hand) perfect equilibrium of agent normal form. In Figure 3.11 we have reduced this trembling to normal form.

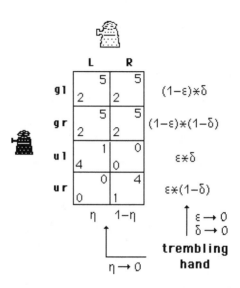

Figure 3.11. How to tremble in normal form

Of course, in this manner of play the second agent does not come into the game at all. The equilibrium choice **r** can nevertheless be interpreted as a belief of the player dalek 2 that the game, should it ever reach the two-element information set, is more likely to be continued at the right-hand decision node.[18]

Yet can this belief be maintained when considered from all points of view? This is doubtful, at least if one, instead of always looking

---

[17]It again suffices to define a threefold sequence of trembling strategies that converge to the corresponding equilibrium.

[18]Here we are discussing in principle the second fundamental refinement of the *equilibrium concept*. In a *sequential equilibrium* (and in its precursor in the history of refinements, the *perfect Bayesian equilibrium*) strategies and beliefs are consistently linked for every player. Thereby beliefs evaluate the realization of corresponding strategic histories, and strategies are based as best replies to conjectures about incomprehensible moves by an opponent.

backward, adopts a point of view of things that could be referred to as forward induction.[19] *Today is a result of yesterday. What this one desired we must discover if we wish to know what that one desires.*

What does dalek 1 desire? If he should actually have a second turn, then a move to the right would bring less utility than the choice of **g** at the root of the game. Thus the belief of player dalek 2 should be revised so that the dalek game, should it ever reach the two-element information set, finds its continuation at the left decision node. Following this argumentation we finally should reject the (trembling hand) perfect equilibrium (**gr**, **R**).[20]

Since neither backward induction nor the trembling hand can protect us from questionable beliefs, we should expand our set of tips to include an additional commandment.

Expanded (Extensive) Tips for Glass Bead Players

1. *Always look backward to plan ahead.*

2. *Consider the impossible; it will exclude itself of its own accord.*

3. *Look ahead if you are relying on beliefs.*

The art of belief is of decisive importance in the following game tree. Originally, *Selten's horse*[21] was sent into the ring as a didactic circus performer [**85**] to demonstrate that not all subgame perfect Nash equilibria are reasonable. In [**53**] the old nag, bound with enormous chains, must even withstand a great leap forward larded with formal obstacles. Only one of the equilibria does not recoil before the hurdle of sequential refinements. Nevertheless, before we establish in great detail the umpire's decision, we would like to carry out a somewhat mythologically tinged interpretation of the horse game.[22]

In the tenth year of the Trojan War the Achaean Epeius, master carpenter, pugilist, and official water-bearer to the house of Atreus,

---

[19] Originally, we had brought Kierkegaard into play as motivation for backward induction. In the matter of forward induction we shall be no less alert and refer to Heinrich Heine's *French Affairs*

[20] In an extensive game every (trembling hand) perfect equilibrium is also sequential.

[21] The reason for the original nomenclature was probably the equine appearance of the game tree depicted in Figure 3.12.

[22] Which admittedly has grown up on our heap of horse manure.

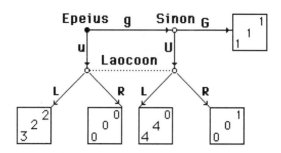

Figure 3.12. Selten's horse with mythological embellishments

created a gigantic, hollow, wooden horse. While the masses of Greek troops hurriedly (and only apparently) abandoned their siege and sailed away, fifty chosen warriors concealed themselves in the horse's interior and waited.

Pretending to be a Greek deserter with a cock and bull story about having been chosen as a sacrifice to placate the winds, Sinon, Odysseus's first cousin, brought the exultant Trojans the news of the departure of the Greeks. The wooden horse had, he said, been left behind by the departing army as a gift dedicated to Pallas Athena. Were it to be brought within the Trojan walls, the city would become impregnable.

Cassandra announced that the horse contained armed soldiers, and of course, no one believed her. She was supported in her claim by Laocoon, priest of Apollo, who exclaimed (according to Virgil, and therefore in Latin) "Quidquid id est, timeo Danaos et dona ferentis," or, if you prefer Robert Graves's rendering, "You fools, never trust a Greek even if he brings you gifts!" Laocoon thereupon deftly drew the game tree of Figure 3.12 in the sand. Out of his sight[23] the situation develops as follows.

If Epeius indeed built the horse—obeying divine inspiration—as a dedicatory gift (move **u**), then Sinon speaks the truth. On the other hand, if it was a question of carrying out an assigned task (Epeius's move is **g**), then the horse is either a gift of the Greeks and the

---

[23]We are describing the game from Laocoon's point of view. Epeius and Sinon move only in Laocoon's imagination, though they thereby nonetheless follow the game-theoretic principles.

defector is lying (move **U**) or a monument by which the Trojan war is to be remembered. In the latter case Sinon would have had no reason to tell the tale of the dedicatory gift (move **G**).

If Laocoon has a turn, he has two options: to approve of bringing the horse into the city (move **L**) or to disapprove (move **R**). Since he does not know the history of the game prior to his information set, he has to rely on reasonable beliefs. Moreover, the concept of a sequential equilibrium makes it necessary that Laocoon's beliefs be stated even if the game never comes around to his turn.

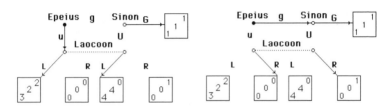

failed in the face of the sequential hurdle mastered the sequential hurdle

Figure 3.13. The game-theoretic great leap forward

In Figure 3.13 we have shown the results of the game-theoretic great leap forward. In order to evaluate them in greater detail we shall be forced to introduce another notation for the above equilibria. Our plan is to make possible an immediate expansion of the subgame perfection argument to game trees that in the usual sense have no subgame trees.

Kreps and Wilson proposed in [**53**] a combination of strategies and beliefs that they called *assessment*. Such an assessment makes it possible to bring in information sets that consists of several decision nodes as the starting point for the game's further development.

In Figure 3.14 we have extended the strategic equilibrium by the requisite beliefs. Epeius and Sinon know with complete certainty where they are when it is their turn. Laocoon's belief, on the other hand, corresponds to a probability distribution over the elements of his multinode information set. In our case Laocoon believes that he

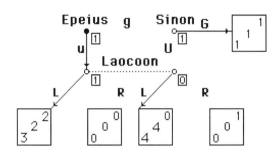

$\boxed{\mu}$ ···· player's belief to find himself at this node
if it is his turn to move again

Figure 3.14. From belief to the reasons for failure

finds himself at the left decision node with probability 1 if it is his turn to move again.

Observe that this belief is doubly compatible with the chosen strategic equilibrium. Firstly, it can be derived directly from the opponents' moves **u** and **G** whose playing (based on imperfect information) in fact does not correspond to Laocoon's observations, though it does to his expectations. Then the same belief perfectly establishes (in the sense of maximizing utility) Laocoon's move **L**.

Epeius, too, has no reason to regret his choice of move **u** if Sinon plays **G** and Laocoon **L** according to his expectations.

The drawback and thereby the grounds for failure in the face of the sequential hurdle are hidden elsewhere. Although the course of the game depicted in Figure 3.14 excludes Sinon, his choice of move does not agree with his expectation as to Laocoon's move. His utility would decrease to four units if on his turn he were to choose move **U** instead of **G**. Now the whole tree—as in Figure 3.15—is razed to the ground.

Namely, if Sinon changes to **U**, then Epeius can by choosing **g** obtain an advantage in utility (case (b) in Figure 3.15). However, this makes Laocoon's belief appear inconsistent. He should now assume that the game continues in the right node of his multinode

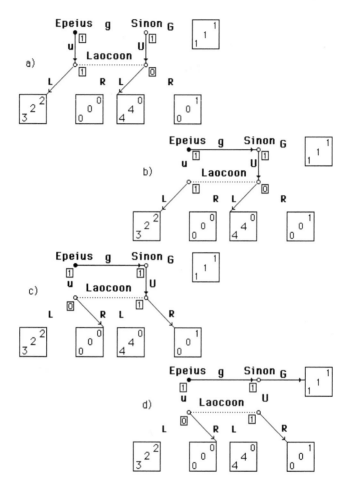

Figure 3.15. No stone rests on another

information set. The altered belief justifies (case (c) in Figure 3.15) Laocoon's change of move to **R**. Finally, all that remains for Sinon is another change to **G** (case (d) in Figure 3.15). The resulting strategic equilibrium together with Laocoon's revised belief withstands the game-theoretic great leap forward.

To prove this, a mathematically educated umpire of leaping must be prepared to provide a sequence of mixed assessments that converge to the given assessment shown in Figure 3.15 under case (d). Here it is required of the elements of this sequence only that the beliefs can be derived from the behavioral strategies in the Bayesian sense.[24] In the limit every strategy must additionally be based on fundamental beliefs in the sense of the best reply.

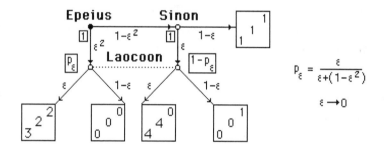

Figure 3.16. The consistency of a sequential equilibrium

But enough of formal incantations! The most fascinating questions of interactive decision theory already appear, namely in games that because of their perfect information seem rather simply constructed. An example of this is the last attraction in our bestiary: Rosenthal's *centipede* [**79**].

If one were to take the name of this game literally—which in the case of most game-theoretic appellations should be avoided at all costs—then one might expect the game tree to take the form of a gigantic beanpole.

---

[24]This follows, for example, in Figure 3.16 by calculating the probability p_ϵ according to the method of calculation given by Bayes's theorem. Here p_ϵ denotes the probability that the game—under the condition that Laocoon gets a turn—has reached the left node of his information set. If Laocoon expects that Epeius and Sinon move downward with respective probabilities ϵ^2 and ϵ, then the required event—that Laocoon's information set be reached in the course of the game—occurs with probability $\epsilon^2 + \epsilon(1 - \epsilon^2)$. According to Bayes's theorem, we thereby have $p_\epsilon = \epsilon/[\epsilon + (1 - \epsilon^2)]$.

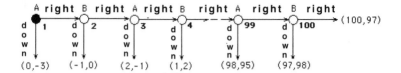

Figure 3.17.   The centipede game

In Figure 3.17 Messieurs A and B alternate turns for as long as the game lasts. The game can always be ended by a move *down* or continued with a move to the *right*. If the downward exit is never chosen, then the game ends with the hundredth move to the right.

At the nth decision node the person whose turn it is (A if n is odd, B if n is even) looks back on a more or less lengthy history $R(n-1)$ that consists of $n-1$ repetitions of the action *right*. If every possible history were enlarged by the action *down* and additionally the history $R(99)$ by the action *right*, then one would have produced all the paths through the game tree that lead to an outcome of the game.

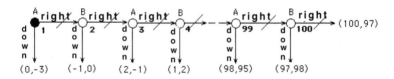

Figure 3.18.   Back induction in the centipede game

It is now not particularly difficult to determine the preferences of both players. For $n \leq 98$ the player whose turn it is at the nth decision node prefers the path $(R(n+1),\ down)$ to the path $(R(n-1),\ down)$ and to the former in turn to the path $(R(n), down)$. While at the end player A would order $(R(99),\ right)$ ahead of $(R(98),\ down)$ and the latter path ahead of $(R(99),\ down)$, his opponent B would incontestably find $(R(99),\ down)$ better than $(R(99),\ right)$.

Now, nothing stands in the way of a simple backward induction—as depicted Figure 3.18. In the subgame perfect equilibrium, the move will be downward, irrespective of the history.[25]

The subgame perfect equilibrium path in Figure 3.19 that results from these strategic considerations is furthermore common to all other Nash equilibria of the centipede game.

Figure 3.19.   The subgame perfect equilibrium path

However, in such a casually produced solution are hidden many logical traps. In Chapter 6 we shall have many problems with these paradoxes of backward induction.

---

[25]Here we are discussing yet again the tiresome strategic concept that often demands that the "impossible plan" be forged. Under the name *counterfactuals* such decisions against the course of things have shaped logical arguments into something much more slippery.

Chapter 4

Games Against Time

> He thought he saw a Chapter on
>   A differential game:
> He looked again, and found it was
>   A Long Prevailing Shame.
> "A lot of reference," he said,
>   "But what's about my name?"
> **—Alexander Mehlmann,** *The Mad Reviewers Song*

At the border between game theory and classical applied mathematics there arose in the 1950s the theory of differential games, initially completely the work of a single individual. The concepts and ideas that Rufus Isaacs [**44**] made use of found (often under the names of others) their expression in the field of optimal control theory that was developing in parallel.

For those in the main currents of game theory the role of this theory seemed to be that of a complex and obscure collection of special cases. This accusation was partially directed at the fight and pursuit situations that stood at the center of these investigations.

In contrast to the conflict situations presented thus far, differential games stress the role of time. However, before we pursue this important influence with the help of two literary conflicts, we shall—with the help of the belligerent theory of duels—say farewell[1] to the classical zero-sum games.

4.1. Duels and Other Affairs of Honor

> Am nächsten Tag steht man befrackt in Tann
> Freds Kraftblick lässt des Gegners Schuss versagen.
> Er selbst trifft ihn am Halse überm Kragen.
> (Ein Kindermädchen trauert in Lausanne.)[2]
> —**Ludwig Rubiner et al.**, *The Duel*

In the Western *Unforgiven* Clint Eastwood plays the gunman William Munny, who at the film's dramatic climax kills five men who have drawn their guns at him. When the dust settles, Beauchamp, a witness to the dispute, asks, "Who'd you kill first? When confronted by superior numbers, an experienced gunfighter will always fire on the best shot first." To this Munny replies drily, "I was lucky in the order. But I've always been lucky when it comes to killin' folks."

Later, in connection with the three-person game truel, we shall advocate Beauchamp's point of view. In a two-person game, or duel, the question of order is beside the point, since it is always clear at whom one is supposed to shoot. Instead of order, here it is the issue of timing that hangs in the balance. Thus the duel belongs to the category of so-called timing games.

In accordance with the traditional rules of the (mathematical) two-person model, the opponents approach each other from an initial distance of A paces. We denote the first duelist's probability of hitting his opponent by $p(x)$ and that of the second by $q(x)$, where the distance between the two adversaries has decreased to x. Both probability functions increase as x approaches zero, at which point both become complete certainties.[3]

---

[1]These mathematical dinosaurs have already completely disappeared from the habitat of game theory and pursue their mischief only in textbooks on linear programming.

[2]They stand among the pines at break of day./ Fred's fell glance thwarts his adversary's bullet./ Then he takes aim and shoots him through the gullet./ (A chambermaid is mourning in Calais.) [Translation by David Kramer.]

[3]Since these functions are strictly monotonic, $x > y$ implies $p(x) < p(y)$ and $q(x) < q(y)$.

If we place a value of $+1$ on sole survivorship,[4] -1 on a sole close encounter with the Angel of Death, and 0 for the other eventualities, namely, that both parties survive or both perish, and if we further assume that each player has but a single bullet at his or her disposal, which when fired issues a loud report,[5] then we can calculate $N_1(x, y)$, the utility that accrues to the first duelist if x is the distance from which the first duelist fires and y the corresponding distance for his opponent, as follows.

For $x > y$ the first duelist will survive the second only if his shot (with probability $p(x)$) is a hit. If he misses (which he does with probability $1 - p(x)$), then without fear of reprisal his opponent can reduce the distance y for his shot to 0 and shoot with probability $q(0) = 1$ of success. Conversely, if $y > x$, then the first duelist will survive his opponent with probability $1 - q(y)$. If the duelists shoot simultaneously, then the first will survive his opponent only if his shot hits the mark without the opponent's shot hitting him. We therefore have

$$N_1(x, y) = \begin{cases} 2p(x) - 1 & \text{if } x > y, \\ p(x) - q(x) & \text{if } x = y, \\ 1 - 2q(y) & \text{if } y > x. \end{cases}$$

In the zero-sum game *duel* the second duelist will always strive to minimize this utility function. To this end he fires at the distance $\hat{y}(x)$ for which

$$N_1(x, \hat{y}(x)) = \min_{0 \le y \le A} N_1(x, y).$$

Now, the accuracy of the first duelist increases the longer he waits (and thus the closer he gets). Thus in no case should the distance $\hat{y}(x)$ exceed x. Let d^* denote the unique distance for which $p(d^*) + q(d^*) = 1$. Then the recipe for minimizing the utility $N_1(x, y)$ is[6]

---

[4] In the interest of maintaining our PG-13 rating we shall make every attempt to depict in the sequel less bloodthirsty scenarios. In Shubik [**88**] the players are content to throw darts at balloons in place of human opponents.

[5] A so-called noisy duel, which has a considerably simpler method of solution than the other variants celebrated in Dresher [**23**] or Karlin [**47**].

[6] This corresponds more closely to a showdown on the streets of Tombstone than to a classical duel. If the minimizing duelist can tell (perhaps by a flicker in his opponent's eye) that his adversary is planning to fire from distance x, then he will also attempt to fire (and fire first) at distance x, provided that the distance d^* has been reached or passed.

$$\hat{y}(x) = \begin{cases} x & \text{if } x \le d^*, \\ 0 & \text{if } x > d^*. \end{cases}$$

The first duelist must content himself with a utility at distance x of

$$N_1(x, \hat{y}(x)) = \begin{cases} 1 - 2q(x) & \text{if } x \le d^*, \\ 2p(x) - 1 & \text{if } x \ge d^*. \end{cases}$$

Nevertheless, with a suitable choice of shooting distance he can maximize this value.

The *maximin* value of the utility function $N_1(x, y)$ is given by

$$\max_{0 \le x \le A} \min_{0 \le y \le A} N_1(x, y) = N_1(d^*, d^*) = p(d^*) - q(d^*).$$

The strategy pair $(x = d^*, y = d^*)$ is thereby the unique *saddle point* of the duel game, since it satisfies the following *saddle point property*:[7]

$$\max_{0 \le x \le A} \min_{0 \le y \le A} N_1(x, y) = \min_{0 \le y \le A} \max_{0 \le y \le A} N_1(x, y) = N_1(d^*, d^*).$$

In this case the *maximin* is equal to the *minimax*.[8] Therefore, the value $p(d^*) - q(d^*)$, which represents both a win for the first duelist and a loss for the second, is called the *value of the game*.

Thus does the exotic flower of paradox eke out a poor existence in the barren soil of zero-sum theory. However, the introduction of just one more gunslinger into the game suffices to alter the situation profoundly.

In a *truel* there are now three opponents facing off, each equipped with an infinite supply of ammunition.[9] Each "truelist" attempts to survive the three-way encounter.[10] We shall give our truelists the names of the three greatest actors in Western films: *John Wayne*,

---

[7]In a zero-sum game every Nash equilibrium has the saddle point property. Conversely, every saddle point is a Nash equilibrium.

[8]With the agreement between these two values John von Neumann has also demonstrated—in his famous minimax theorem—the existence of a game value for every finite two-person zero-sum game.

[9]In the literature (see Kilgour [48]) this special case is called an infinite truel.

[10]If each participant hopes to be the only survivor of the contest, then one speaks of an unambiguously antagonistic truel. If there is at least one truelist who does not care whether he alone survives or whether others survive with him, then the truel has cooperative moves.

Clint Eastwood, and *Randolph Scott*. John, let us suppose, is the best shot, followed by Clint, and then Randolph. Our truelists will stand at the vertices of an equilateral triangle, and these positions will remain fixed during the entire exchange of gunfire. Thus the probability functions for the three may be reduced to constants $j > c > r$.

The truelists begin by drawing lots for the order of firing,[11] and this order will be strictly maintained for the duration of the truel.[12]

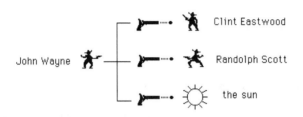

Figure 4.1.   John Wayne's truel strategy

In Figure 4.1 we have shown the possible strategies for John Wayne. When it is John's turn to shoot, if both his opponents are still among the living and if he has decided once and for all to shoot in such situations at the best marksman, then he will shoot at Clint and hit him with probability j.

To be sure, John would have just as good a chance of eliminating Randolph, but in looking ahead, it becomes clear that John's probability of winning the truel would diminish in this case, since as the shooting progresses John would become the target of an opponent who is a better shot.

In an unambiguously antagonistic truel in which the only permitted targets are other players, it turns out that the strategy of shooting at the strongest opponent is always the optimal one. In Figure 4.2 we have depicted the resulting (unique) Nash equilibrium.

So, is John sitting pretty? Although John is by far the best marksman, he may find himself, as far as the probability of survival

---

[11] There are six possible firing orders in all.

[12] In the event that one of the gunmen bites the dust, his turn will simply be omitted.

is concerned, trailing behind Randolph. This paradoxical outcome was first described by Shubik in [**87**].

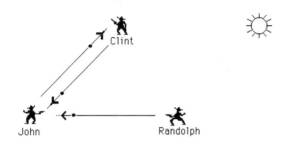

Figure 4.2.   The weak are more likely to survive—Shubik's solution

For truels with cooperative moves Gardner [**31**] describes an unorthodox (additional) equilibrium in which as long as all three opponents are still alive, the third player shoots into the air instead of at one of his adversaries. The circumstances under which a voluntary waste of a shot is the optimal response to the opponents' strategies, as depicted in Figure 4.3, depend—according to Kilgours's comprehensive analysis of truels [**48**]—both on the order of shooting and on the skill of the second-best shooter.

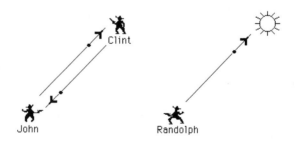

Figure 4.3.   The third man—Gardner's solution

Namely, if Clint is not all that good a marksman, then Randolph, an even more pathetic shot, will always (regardless of the order of shooting) take aim at John. However, if Clint is a crack shot and has his turn directly after Randolph's, then Randolph will continually shoot at the sun until one of his stronger opponents is eliminated. At that point Randolph can take aim at the survivor and let him have it.

Donald Knuth, the creator of the magnificent typesetting language TeX (in whose refined offshoot LaTeX this book has been set), provides in [**50**] a truly pacifistic finishing touch[13] to the most cooperative of all possible truels.[14]

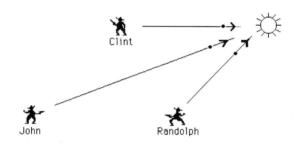

Figure 4.4. A truel under the sun—Knuth's solution

4.2. The Curse of Irreversibility

D'année en année, cependant, son petit champ se rétrécit, et, lorsqu'il survient une épidémie, il ne sait pas s'il doit se réjouir des décès ou s'affliger des sépultures. –Vous vous nourrissez des morts, Lestiboudois! lui dit enfin, un jour, M. le curé.[15]

—**Gustave Flaubert,** *Madame Bovary*

In *Madame Bovary* Gustave Flaubert introduces as filler material the figure of the sexton Lestiboudois. Alongside his activities as gravedigger, this upright man uses the fallow part of the cemetery to grow

---

[13] We doubt whether the Pareto-efficient equilibrium shown in Figure 4.4 would suffice as a game-theoretic explanation for the origin of sunspots.

[14] Each player is indifferent as to whether he survives alone, or with one or even two others.

[15] Nevertheless, his little field grows smaller every year, and when there is an epidemic he doesn't know whether to rejoice in the deaths or lament the space taken by the new graves. "You are feeding on the dead, Lestiboudois!" Monsieur le curé told him, one day. [Translation by Francis Steegmuller.]

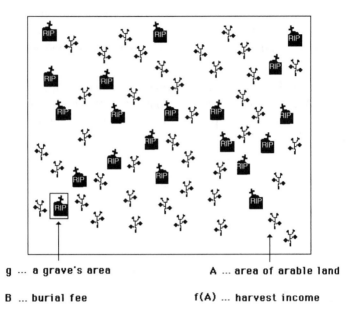

g ... a grave's area   A ... area of arable land

B ... burial fee   f(A) ... harvest income

Figure 4.5. Lestiboudois's graveyard and potato field

potatoes. Before he knows it, he is caught up in a game against nature, which among other things determines within a certain range $[\underline{m}, \overline{m}]$ the death rate $m(t)$ of the parish.

On the other hand, Lestiboudois seems limited in the amount of leeway at his disposal. Yet one should not completely underestimate the influence of a sexton. We therefore would like—for the sake of the model—to assume that it is possible for Lestiboudois, through rumors that lives of certain of the deceased were in no way pleasing to God, to reduce to zero the number of burials (within a certain period of time). To determine the present value of Lestiboudois's future utility stream it seems advisable to pose, following Thépot [**90**], the maximization problem

$$\max_{0 \le \omega(t) \le 1} \int_0^\infty e^{-rt}[f\big(A(t)\big) + B\omega(t)m(t)] \, dt$$

under the dynamic supplementary condition

$$\dot{A}(t) = -g\omega(t)m(t), \qquad A(0) = A_0.$$

Now, Lestiboudois controls the number of burials by means of the acceptance rate $0 \leq \omega(t) \leq 1$. This decision influences the amount of land that is available to him to place under cultivation, whose decrease over time is expressed by the above differential equation. Here $\dot{A}$ denotes the derivative of A with respect to time t.

At every moment in time Lestiboudois's utility is determined by the amount of his potato harvest and his income from burials. If these values are exponentially weighted with the discount rate r and accumulated over an infinite time horizon, then one obtains the objective function to be maximized.

For the case of a strictly concave function[16] $f(A)$, a certain value $\hat{A}$ can be expressed by the equation

$$f'(\hat{A}) = \frac{rB}{g}$$

in such a way that for an area of arable land A_0 at the beginning of the planning period that exceeds this value, Lestiboudois's best reply to nature's move consists in not missing the opportunity for any burials for the time being. This decision is irreversible; it continually reduces the area available for planting and finally forces Lestiboudois in the case where only the amount $\hat{A}$ of land is available to him for planting to neglect his gravedigging duties completely. How does this time-dependent best reply come into being? Lestiboudois must certainly make his decisions under the dynamic condition of the reduction in the size of his potato field. In accordance with the recommendations of the Pontryagin maximum principle one can nonetheless fall back instead on the solution of the following infinite system of static game situations:[17]

$$\max_{0 \leq \omega(t) \leq 1} \left\{ f\big(A(t)\big) + B\omega(t)m(t) + \lambda(t)[-g\omega(t)m(t)] \right\}.$$

This solution is determined at every point in time t by

$$\hat{\omega}(t) = \begin{cases} 1 & \text{if } \lambda(t) < \frac{B}{g}, \\ 0 & \text{otherwise,} \end{cases}$$

---

[16]That is, a function whose first derivative in the argument A is strictly positive and whose second derivative is strictly negative.

[17]Which are essentially one-person games.

where the valuation of the dynamic reduction of the arable area occurs according to the shadow price

$$\dot{\lambda}(t) = r\lambda(t) - f'\big(A(t)\big).$$

A compendium of essential methods for the maximum principle is provided by Feichtinger and Hartl [27]. In the area of differential games we refer to Mehlmann [59]. In the following section we shall turn our attention to a classical—in the literal sense—application[18] of the theory of differential games.

4.3. Mephisto's Joust with Doctor Faust

> Werd ich zum Augenblicke sagen:
> Verweile doch! du bist so schön!
> Dann magst du mich in Fesseln schlagen,
> Dann will ich gern zugrunde gehn![19]
> —**Johann Wolfgang von Goethe**, *Faust*

The original Faust motif begins with a limited-time closed devil's pact between the fisher of souls and the old academic. With Goethe this pact experiences a transformation into a wager whose essential criterion is the determination of the point in time at which Faust will lose his soul.

Mephisto conjectures that the moment of truth can be realized only by means of seductive machinations,[20] and he estimates the probability of this (for him) totally pleasant outcome as being directly proportional to the current seduction intensity $u_1(t)$, that is,

$$\dot{x}_1 = c_1 u_1 (1 - x_1),$$

where c_1 is a constant and the initial value is given by $x_1(0) = 0$.

---

[18]It deals with one of the few obsessions that the author of these lines has publicly and on many occasions (among others in [61], [28], and [60]) confessed to.

[19]If to the moment I should say:/ Abide, you are so fair—/ Put me in fetters on that day,/ I *wish* to perish then, I swear. [Translation by Walter Kaufmann.]

[20]Ein solcher Auftrag schreckt mich nicht/ Mit solchen Schätzen kann ich dienen; *Faust 1*, ii, 1688–1689. (Such a commission scares me not,/ With such things I can wait on you.) [Translation by Walter Kaufmann.]

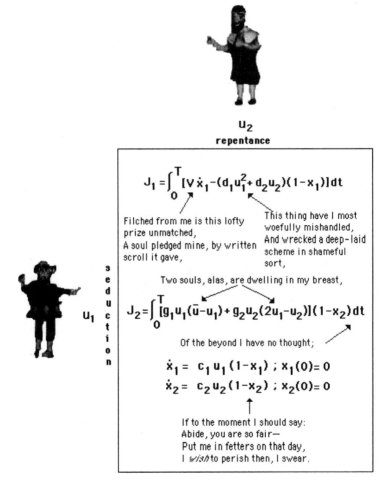

Figure 4.6. A wager with the devil

Faust, on the other hand, doubts[21] Mephisto's point of view. He is aware that the moment of truth can be reached only by (active)

---

[21]Was willst du armer Teufel geben?/ Ward eines Menschen Geist, in seinem hohen Streben,/ Von deinesgleichen je gefaßt; *Faust* 1, ii, 1675–1677. (What would you, wretched Devil, offer?/ Was ever a man's spirit in its noble striving/ Grasped by your like, devilish scoffer?) [Translation by Walter Kaufmann.]

repentance, that is, in the mirror image of the devil's formula

$$\dot{x}_2 = c_2 u_2(1 - x_2),$$

where c_2 is a constant, the initial value is given by $x_2(0) = 0$, and $u_2(t)$ is the instantaneous repentance.

As depicted[22] in Figure 4.6, Mephisto's expected payoff function J_1 consists of two components—each weighted with the probability of the event pertaining to it.

If Mephisto wins the wager at time t, then he receives the equivalent value[23] V for Faust's soul. If the moment of truth has not yet been reached, then the poor devil must figure on a squared expenditure[24] $d_1 u_1^2$ and the disutility $d_2 u_2$ due to Faust's repentance.

In contrast to Mephisto, Faust has no expectations of the hereafter.[25] From a Freudian point of view and according to Goethe's interpretation[26] we may associate the motivation and components of the second objective function with the various layers of Faust's soul. The hedonistic id obtains the concave utility $g_1 u_1(\bar{u} - u_1)$ from the present seduction, where $\bar{u}$ stands for the natural bound on his libidinous needs. The moralistic superego can regret at most the current seduction, which can be derived immediately from the second utility term $g_2 u_1(2u_1 - u_2)$.

If we define for each player time-dependent covariables μ_{ij} for $i, j = 1, 2$ as the valuation of the game state x_i over time, then we can form for each time t the following infinite collection of static game

[22]The images of Mephisto and Faust in that figure are photographs of the author's two Czech marionettes, which he has often pressed into service to illustrate his talks on the game-theoretic implications of the Faustian bargain.

[23]Mir ist ein großer, einziger Schatz entwendet:/ Die hohe Seele, die sich mir verpfändet; Faust 2, ii, 11828–11829. (Filched from me is this lofty prize unmatched,/ A soul pledged mine, by written scroll it gave,) [Translation by Walter Kaufmann.]

[24]Ein großer Aufwand, schmählich! ist vertan; Faust 2, v, 11837. (This thing have I most woefully mishandled,/ And wrecked a deep-laid scheme in shameful sort,) [Translation by Philip Wayne.]

[25]Das Drüben kann mich wenig kümmern; Faust 1, ii, 1660. (Of the beyond I have no thought;) [Translation by Walter Kaufmann.]

[26]Zwei Seelen wohnen, ach, in meiner Brust, Faust 1, i. Vor dem Tore. (Two souls, alas, are dwelling in my breast,) [Translation by Walter Kaufmann.]

situations:

$$\max_{u_1}\big\{ V\dot{x}_1 - (d_1 u_1^2 + d_2 u_2)(1 - x_1) + \mu_{11}\dot{x}_1 + \mu_{12}\dot{x}_2 \big\},$$

$$\max_{u_2}\big\{ [g_1 u_1(\bar{u} - u_1) + g_2 u_2(2u_1 - u_2)](1 - x_2) + \mu_{21}\dot{x}_1 + \mu_{22}\dot{x}_2 \big\}.$$

Faust now reacts (in the sense of his best reply) to each diabolical machination u_1 according to[27]

$$\hat{u}_2 = u_1 + \frac{\mu_{22}c_2}{2g_2}.$$

If Mephisto should get mixed up in a simultaneous game against Faust, then his best reply is

$$\hat{u}_1 = \frac{c_1(V + \mu_{11})}{2d_1}.$$

Much more diabolical would be, however, if he—in calculating foresight of the expected reflex from Faust—were to choose the seduction amount

$$\tilde{u}_1 = \frac{c_1(V + \mu_{11}) - d_2 + c_2\mu_{12}(1 - x_2)(1 - x_1)^{-1}}{2d_1},$$

which is favorable at any time.

The strategic pair $(\hat{u}_1, \hat{u}_2)$ represents a Nash equilibrium of the infinite collection of static game situations in simultaneous play. If one assumes, to the contrary, that Faust has the first move, then one obtains $(\tilde{u}_1, \hat{u}_2)$ as an equilibrium. To obtain from this the corresponding equilibria of the differential game one has merely to replace the covariables by solutions of the differential equations derived by the maximum principle. For simultaneous play these equations are as follows:

$$\dot{\mu}_{11} = -(d_1 u_1^2 + d_2 u_2) + c_1 u_1 \mu_{11},$$

$$\dot{\mu}_{22} = -[g_1 u_1(\bar{u} - u_1) + g_2 u_2(2u_1 - u_2)] + c_2 u_2 \mu_{22},$$

$$\mu_{11}(T) = 0,$$

$$\mu_{22}(T) = 0.$$

---

[27]One simply takes the first derivative of the static objective function belonging to Faust with respect to u_1 and sets it equal to zero.

Direct statements about the ending of the devil's wager can be derived only under the assumptions given by Mehlmann and Willing in their mathematical ur-Faust [**61**]:

Faustian–Mephistophelean Consistency Theorem

1. *The more that repentance irritates Mephisto and the higher he sets the expected accumulated seduction $1/c_1$ that would be necessary to produce the moment of truth, the higher must be the price on Faust's soul for the wager to be played out at all.*

2. *The higher Faust's libido is set, the less will seduction put Mephisto out of pocket, or, more precisely, the less will repentance bring satisfaction to Faust, but, on the other hand, the higher will repentance irritate Mephisto, or, more precisely, will seduction entice Faust.*

3. *The evaluation of Faust's utility arising from seduction must come to at least three-fourths of the evaluation of the profit accruing to him through repentance, that is, $4g_1 \geq 3g_2$.*

For the case of a higher evaluation of the utility that Faust obtains from the seduction, the equilibrium interplay of seduction and repentance can be described in Figure 4.7.

Although Faust can win far more from seduction, his equilibrium strategy consists in overrepentance. This seemingly paradoxical behavior can be explained as follows. Since Mephisto is so disturbed by Faust's repentance, he must press for a quick end to the game. He therefore estimates his machinations much too highly and reduces them only when the value[28] V of the soul is not sufficiently tempting.

In reality, therefore, Faust derives no utility from what the devil offers him. Since the disutility from excessive seduction is, moreover, at least twice as great, Faust himself has a great interest in shortening the wager by means of overrepentance. This might also be the reason why in the second part of *Faust* the outcome is so disappointing to

---

[28] Depicted in Figure 4.7 as a point on the dashed line $u_2 - u_1 = 0$.

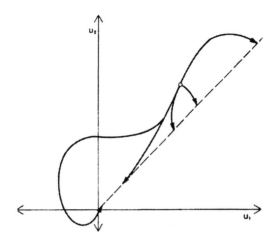

Figure 4.7. A phase diagram of Goethe's *Faust*

Mephisto. He realizes that his view of the game was false[29] and that it was he who was deceived.

In Figure 4.8 is depicted a pendant to Goethe's tragedy. A tenable interpretation of this situation can be made by reference to the original Faust motif—as presented, for example, by Christopher Marlowe to great dramatic effect.

Mephisto forces Faust in a diabolical manner to underrepentance—taking advantage of his stronger evaluation of repentance utility. In contradistinction to the earlier diagram, Faust's total utility is thereby positive, while Mephisto's is diminished, a situation that makes a timely termination of the wager unlikely. Just before time runs out Mephisto for a time records a reduction in seduction intensity, while Faust increases his repentance to tear himself from the devil's clutches.

---

[29] "Ihn sättigt keine Lust, ihm gnügt kein Glück." Faust 2, *v*, line 11587. (Him would no joys content, no fortune please,/ And thus he wooed his changing fantasies.) [Translation by Philip Wayne.]

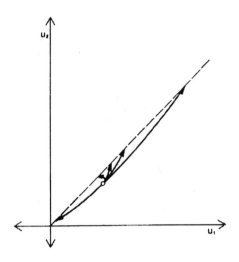

Figure 4.8. A phase diagram for Marlowe's *Faust*

For a relatively high soul value one can observe in Marlowe the lessening of the devil's activities and Faust's all-too-late—and therefore pointless—repentance.[30]

---

[30] "Accursed Faustus, wretch, what hast thou done? I do repent; and yet I do despair." *The Tragical History of Doctor Faustus*, Act V, Scene 14.

Part II

The Myths of Game Theory

Introduction to Part II

Contemporary man has rationalized the myths,
but he has not been able to destroy them.
—**Octavio Paz,** *The Labyrinth of Solitude*

Can one honestly admire the patterns of game theory without recalling the riddles, paradoxes, and myths from which sprang the first strategic lines of reasoning? The collection of models in the following part of the book can be offered as supporting documentation for the cross pollination between the mathematics of conflict and the literary tradition of myths that has been handed down to us. We begin with the peculiar example of a mathematical model that became a narrative myth.

In the sequel we shall be concerning ourselves with paradoxes[31] of backward induction and fundamental questions of game-theoretic scholasticism.

Along the path from knowledge, common to the logical quicksand of counterfactuals we shall make use of literary motifs as guideposts.

---

[31] The following famous quotation comes from **Francisco José de Goya y Lucientes**. It appears engraved on the forty-third plate of *Los Caprichos*: "El sueño de la razón produce monstruos" (The sleep of reason produces monsters). In connection with game theory it has a fascinating converse. The monsters of paradox have more the state of wakefulness than the sleep of reason to thank for their existence.

Finally, this second part, on the myths of game theory, earns a worthy conclusion. In the last chapter a game-theoretic model will attempt a strategic elucidation of the ninety-fifth fable of Hyginus.

Chapter 5

The Prisoner's Dilemma

> Each one of a gang, so placed, is not so much greedy of reward, or
> anxious for escape, as fearful of betrayal. He betrays eagerly and
> early that he may not himself be betrayed.
> — **Edgar Allan Poe,** *The Mystery of Marie Rogét*

Is game theory a hunting ground for truisms? At first glance one
is thoroughly tempted to give in to this impression. The reception
of Albert Tucker's omnipresent anecdote has if nothing else enriched
the rhetorical repertoire of professional boasters with an additional
trump card. The world that up to now was round is no longer a zero-
sum game, and anything that creeps or crawls is at the mercy of the
prisoner's dilemma.

We shall proceed in what follows on a fanciful search for a para-
digm to describe situations in which the temptation to deceive one's
opponent triumphs over one's readiness to trust him.

5.1. Variants That We Knew

What could have possessed the venerable Albert W. Tucker, a Princeton mathematician by trade, when one fine May morning (in the year
1950) he revealed to psychologists at Stanford the secrets of an up-
and-coming young discipline called game theory?

All that he could do with a clear conscience when faced with an audience of specialists in a field foreign to him was to present a strange experimental game into which his colleagues at RAND had initiated him when he was working there as a consultant. Merill Flood and Melvin Dresher had hunted down, to the greater glory of science, two innocent human guinea pigs in a hundred-round battle over a laughable sum. In the course of this legendary conflict the unfortunate combatants anticipated developments that for almost half a century have shaped the development of modern game theory.

In the interplay of cooperation, breach of faith, revenge, and forgiveness a unique result crystallized at the statistical mean—the tendency to continued cooperation, which held fast in gross violation of the solution expected from all sides: the Nash equilibrium.

Tucker had correctly conjectured that this behavior was to be attributed only to the interactive-dynamic character of the (repeated) game. He reduced the time frame to a single confrontation and thereby brought the validity of the theory of his favorite student, Nash, again into alignment. His next variation, however, caused something much more decisive: It transformed a purely schematic numerical example into the living mythology of the prisoner's dilemma.

Like a game of telephone, transformed on each retelling into yet another formulation, Tucker's protean anecdote traveled over the course of time from textbook to textbook—a worthy tradition that willy-nilly we were unable to ignore.

Tucker's Anecdote

Bonnie and Clyde are caught after an attempted bank robbery that went awry and have been placed in the county jail in separate cells. If they do not confess to the crime, the sheriff can prove only illegal possession of a firearm. For such an offense the penalty is three years in jail. If one of the pair remains steadfast and silent but the other confesses, then the one who has confessed, as witness for the prosecution, will spend only one year in the slammer, but his stronger-willed partner will be slapped with a nine-year sentence. If they both confess, then they each draw a seven-year sentence. Presented with this suite of choices, how will they act?

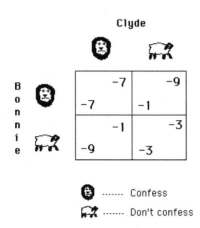

Clyde

 ······· Confess

 ······· Don't confess

Figure 5.1. Tucker's anecdote as a bimatrix game

In Figure 5.1 we have transformed the prison sentences into negative utility values. In accordance with the indication in footnote 3 of Chapter 2, it is a question neither of the origin of our utility scale nor of the size of our unit of utility. We can thus increase all the utility values by the amount 7 and then divide by 4 without altering the essentials of our game.

The game matrix displayed in Figure 5.2 can rightly be called the "mother of all prisoner's dilemmas." Here $-s$ denotes the *sucker's payoff*, while t stands for the payoff to the traitor. For the sentences handed out in Tucker's anecdote one obtains, after the proposed positive linear transformation, $s = \frac{1}{2}$ and $t = \frac{3}{2}$.

One should take note that the two players are unable to communicate with each other (there is no possibility of smuggling secret messages between cells, and they have no access, of course, to cell phones),[1] and thus for each player it is more than doubtful whether the other will refrain from making a statement. In the language of

---

[1] Cells are also involved in the only verifiable biological occurrence of the prisoner's dilemma. Turner and Lin Chao [**92**] have shown that certain viruses that infect and then reproduce in the same host cell are engaged in a survival-of-the-fittest-driven prisoner's dilemma.

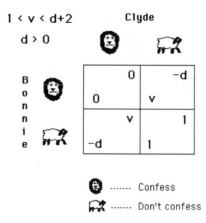

Figure 5.2.   The mother of all prisoner's dilemmas

game theory, one says that there can be neither an exchange of information nor a binding agreement between the players.

The bimatrix of the prisoner's dilemma is known to possess a unique Nash equilibrium. Namely, if we assume that Bonnie confesses, then it would be pure madness for Clyde not to confess as well, since in this case he reduces his utility, the sucker's payoff, from 0 to $-s$. Exactly the same argument can be applied to the case of a confession by Clyde. Nevertheless, there is exactly one outcome that would bring both players more than what they achieve at the equilibrium. If they both decline to confess, they could increase their winnings from 0 to 1. Sad to say, this mode of play does not lead to an equilibrium. If Bonnie does not confess, then Clyde would doubtless at once confess, since this increases his winnings from 1 to t, the traitor's payoff. Although the cooperative mode of play of not confessing thereby dominates the equilibrium when measured by the standard of payoff, the noncooperative theory of normal form games cannot select this outcome as a solution of the prisoner's dilemma.

This fact has mistakenly been judged a serious shortcoming in the solution theory of non-zero-sum games. In what follows we are going to discuss two of the multitude of approaches that can assist us in overcoming this alleged defect. Common to all such approaches is the rejection of the static (that is, single-period) form, which (correctly and in conformity with the rules of the normal form game) assumes that each player is allowed only one (final) move.

The plot of the prisoner's dilemma has, moreover, been the cause of justified criticism. It rejects the obviously (social and criminal) previous experience of the two accused parties. What if Bonnie and Clyde were romantically involved? Or might they have enough Sicilian ancestry to know the meaning of *omertà*? Questions on top of questions that not least have awakened the demand for sound variants of the behind-bars melodrama.

So we may thank Rapoport [**74**], among others, for the unalloyed artistic pleasure of a prisoner's dilemma for opera lovers.

Rapoport's *Tosca* Paraphrase

In a desperate attempt to rescue her lover Cavaradossi, who has been condemned to die before a firing squad, Tosca makes a fatal pact with the henchman Scarpia. She is prepared to give herself to him if he agrees, before the encounter, that the execution will be carried out with blank cartridges. The Charybdis of a prisoner's dilemma draws Tosca and Scarpia into its vortex, not, of course, without giving each of them the opportunity to sing a final aria. Each then breaks the agreement by choosing his or her strictly dominant strategy. Thus Scarpia secretly countermands the order to switch to blanks, while Tosca, for her part, stabs her love-crazed adversary with a knife that seems somehow to have been left on stage by an absent-minded property mistress.

The values postulated by Rapoport for the sucker's payoff and that of the traitor are in no way inferior in plausibility to the opera's libretto.

A more unusual solution to the prisoner's dilemma is suggested by one of Gregor von Rezzori's most beautiful Maghrebinian tales.

The Maghrebinian Prisoner's Dilemma

In Maghrebinia criminals who confess are punished twice as harshly, "since to the insolence of their crime against the laws they add the shamelessness of admitting to it" ([**76**], *p. 156*). *It therefore does not pay to offer a confession. Namely, if one doubles the penalties shown in Figure 5.1 for those who confess, then one obtains (as is apparent in Figure 5.3) the unique Nash equilibrium* (do not confess, do not confess).

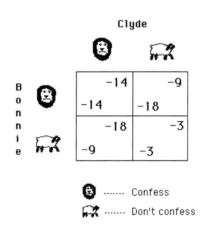

Figure 5.3. Maghrebinian prisoner's dilemma

5.2. The Tournament of the Automata

Axelrod [**5**] analyzed an interesting approach to the solution of the prisoner's dilemma that consists in allowing a possible repetition of the game. Mathematically, this means that there exists a positive number w^{n-1} that represents the probability that Bonnie and Clyde end up for the nth time in the same (bad) situation.

If this is the case, then they can bring into their deliberations both positive and negative experiences of the previous outcomes and thereby develop strategies based on this information.

This brings about a return to the descriptive roots of the prisoner's dilemma. Instead of facing two alternatives in a static game,

Bonnie now has at her disposal strategies equipped with a memory for a game that (in the case of a strictly positive w) never ends.

To obtain an overview of suitable strategies that have it in them to represent cooperation as a desirable goal in the prisoner's dilemma, Axelrod advertised a computer tournament in which such strategies could compete against one another.[2] Strategies, which were submitted in the form of short or not-so-short computer programs, had the possibility of entering into the lists against themselves and against all the other programs. The winners were to be those strategies that emerged with the highest number of points after all of the jousts were completed.

The number of iterations of the single-stage dilemma that was planned for each encounter was concealed from all the participants. While Axelrod originally had held out to the participants the prospect of two hundred rounds in his tournament, he eventually shaped the duel into an infinitely repeatable game. To take the wind out of possible protests against this blatant change in the rules, he took care that the expected number of repetitions corresponded at least to the originally proposed number of rounds.

The Teachings of Axelrod's Tournament

Axelrod's tournament is exemplary for a number of reasons. It demonstrates the following:

1. *How in a game (through the principle of repeatability) learning behavior and memory can arise.*

2. *How strategies, even in an infinitely repeated game, require information based only on finitely many environmental conditions to react successfully.*

3. *That—as any football coach will tell you—"Tactics are precisely what your opponent allows you to play."*

4. *That success depends on the number and type of counterstrategies in an evolutionary sense.[3]*

---

[2] The paths to fame are often tortuous, and often strange. Axelrod is probably the first modern knight of science who achieved renown by way of a tournament.

[3] Here the revenge of the mutants comes to mind.

Tit-for-tat, a strategy that was sent in by Anatol Rapoport, achieved victory despite its simplicity. Its strategic program can be expressed thus: *Cooperate in the first round (that is, do not confess), and in every subsequent round do exactly what your opponent did in the previous round.*

A picture, as they say, is worth a thousand words. For this reason we have decided on the following graphic (even though mathematically more demanding) description of the *tit-for-tat* strategy.

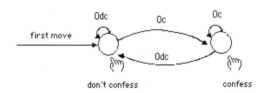

Figure 5.4. Tit-for-tat as a Moore machine

The first move initializes the strategic behavior of the player. Transitions from one active behavioral state into another are triggered by the opponent's action in the previous round. The available information is hereby either **Oc** (opponent confessed) or **Odc** (opponent did not confess). The player has now found, in the form of a finite automaton, an ideal representative for a joust in a computer tournament.

No one can explain better than opinion pollsters, meteorologists, and careless lovers the reasons for the arrival of an unexpected event. Axelrod demonstrated his abilities as a Monday-morning quarterback when in his evaluation of tit-for-tat he discovered (after the fact) several unmistakable characteristics of the winners. The first of these characteristics was *transparency*, where for us in this connection this term seems to mean *recognizability* by all conceivable opponents. In contrast to the nasty *mean tit-for-tat*, tit-for-tat is *nice*, since it cooperates in the first round.

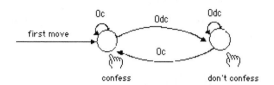

Figure 5.5. Mean tit-for-tat as a Moore machine

Tit-for-tat has the property that it is *not vindictive* to nice strategies such as *liverwurst*.[4]

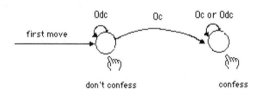

Figure 5.6. Liverwurst as a Moore machine

Finally, tit-for-tat is *provocable*. Indeed, after a comparison with the likewise provocable strategy *T-for-two*[5] one might consider it more *hot-tempered* than provocable.

Game theorists generally consider this latter characteristic as the obvious weakness of the otherwise successful rules of the game in the tournament. It leads to punishing a single slip mercilessly. Thus, for example, a test of strength between tit-for-tat and its evil twin leads after the second round to an unalterable and asynchronous cycling between confession and cooperation. T-for-two and mean tit-for-tat, on the other hand, find themselves after the same period of time on

---

[4]Whose trademark is the feeling of having been offended, and not so much that of being unrelenting, as the common name for this strategy in the literature, *grim*, would lead one to think. (Translator's note: In German the expression *beleidigte Leberwurst spielen*—to play the offended liverwurst—means to be in a huff.)

[5]In all seriousness pronounced *tea for two*. Those practitioners lacking a sense of humor usually call this strategy *tit-for-two-tat*. As we say in German, *Chacun à son goût*.

a course of permanent cooperation. Axelrod's assessment is that T-for-two would have wrested the laurels from tit-for-tat had it taken part in the tournament.

5.3. Foresighted Equilibria

As a generalization of the solution concept of a *nonshortsighted equilibrium* introduced by Brams and Wittman [**12**], Kilgour [**49**] proposes the following procedure:

In order to be able to review the available selection of strategies, every player plays over (in his or her mind) extensive games that permit a deviation from a given pair of strategies (that is, game starting points). Let us assume, for the moment, that Bonnie chooses the outcome (*confess, confess*), which in the sequel we shall denote for the sake of simplicity by its evaluation $(0,0)$, as the starting point for her deliberations.

Bonnie pictures herself in the role of the first player whose turn it is in an extensive game that runs as follows. She has two possibilities: $[M]$, to move away from the starting point $(0,0)$, or $[R]$, to remain at the starting point $(0,0)$.

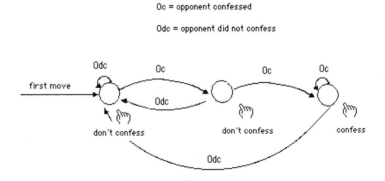

Figure 5.7. T-for-two as a Moore machine

If Bonnie chooses $[M]$, then the new starting point is given by $(-s, t)$ (since Bonnie can move further only in the same column of the bimatrix), and Clyde now has the decision whether to be content with $(-s, t)$ or to move away (in that he moves to the next starting point appearing in the same row as $(-s, t)$).

We shall now put a time limit on this extensive game, by assuming that after the kth decision $[M]$ the game tree of the extensive game is truncated. The game is now at an end if the player whose turn it is decides to remain.

For the special case $k = 3$ we have shown in Figure 5.8 the game tree of the extensive game that describes Bonnie's decisions with respect to the position $(0, 0)$.

$$
\begin{array}{lll}
(0, 0) & [R] & \\
\textbf{Bonnie} & \rightarrow & (0, 0) \\
{[M]} \downarrow & & \\
(-s, t) & [R] & \\
\textbf{Clyde} & \rightarrow & (-s, t) \\
{[M]} \downarrow & & \\
(1, 1) & [R] & \\
\textbf{Bonnie} & \rightarrow & (1, 1) \\
{[M]} \downarrow & & \\
(t, -s) & &
\end{array}
$$

Figure 5.8. Bonnie's moves from position $(0, 0)$

This extensive game can be solved relatively easily by introducing backward induction. We begin with the player at his or her last move and decide in favor of the move that obviously offers the greater profit. In our example, this player is Bonnie (Figure 5.9).

$$
\begin{array}{lll}
(1, 1) & [R] & \\
\textbf{Bonnie} & \nrightarrow & (1, 1) \\
{[M]} \downarrow & & \\
(t, -s) & &
\end{array}
$$

Figure 5.9. Bonnie's last move

$(-s, t)$ $[R]$

Clyde $\rightarrow$ $(-s, t)$

$[M]$ ⤏

$(1, 1)$ $[R]$

Bonnie   ⤏ $(1, 1)$

$[M]$ ↓

$(t, -s)$

Figure 5.10.   Clyde's move

Bonnie should choose $[M]$. The move $[R]$ is then to be eliminated. (This is accomplished simply by drawing a line through it.)

At the penultimate stage it is Clyde's move (Figure 5.10).

If Clyde chooses $[R]$, then he can figure on a payoff of t. On the other hand, if he chooses $[M]$, then he reaps the sucker's payoff $-s$, since Bonnie's decision at the last step is fixed at $[M]$. Thus Clyde will choose $[R]$, and we chop off the remaining branch of the game tree.

We have now arrived at the beginning and can assure Bonnie of a payoff of 0 if she chooses $[R]$. But if she should prefer $[M]$, then she will achieve only the value $-s$. Thus for $k = 3$ we have the decision depicted in Figure 5.11.

Bonnie remains at $(0, 0)$. This decision will not change if the number of moves (k, that is), is increased. If Clyde gets to go first, then he, too, will vote for remaining at $(0, 0)$.

If for each given outcome of the game and for both players we obtain a natural number l such that for each $k \geq l$ the game tree analysis recommends remaining at the game's starting point, then we shall speak of a *foresighted* equilibrium point.

To determine whether $(0, 0)$ is the only foresighted equilibrium point for the prisoner's dilemma (which for obvious reasons we hope not to be the case), we must carry out the game tree analysis for the remaining starting points (Figure 5.12).

$(0,0)$ $[R]$

Bonnie $\rightarrow$ $(0,0)$

$[M] \not\downarrow$

$(-s,t)$ $[R]$

Clyde $\rightarrow$ $(-s,t)$

$[M] \not\downarrow$

$(1,1)$ $[R]$

Bonnie $\nrightarrow$ $(1,1)$

$[M] \downarrow$

$(t,-s)$

Figure 5.11.   Bonnie's decision

$(1,1)$ $[R]$

Bonnie $\rightarrow$ $(1,1)$

$[M] \not\downarrow$

$(t,-s)$ $[R]$

Clyde $\nrightarrow$ $(t,-s)$

$[M] \downarrow$

$(0,0)$ $[R]$

Bonnie $\rightarrow$ $(0,0)$

$[M] \not\downarrow$

$(-s,t)$ $\ldots$

$\vdots$

Figure 5.12.   Bonnie remains in position $(1,1)$

After the second move we have reached a game tree that has already been analyzed. By the symmetry of the bimatrix, for the starting point $(1,1)$ (that is, (*don't confess, don't confess*)), each player has the same pattern of decisions.

The point $(1, 1)$ is also a foresighted equilibrium point. Since it dominates the foresighted starting point $(0, 0)$ in terms of its value, it can be considered the only (foresighted) solution of the prisoner's dilemma.

Finally, how do we evaluate the two remaining starting points? Bonnie would remain at $(t, -s)$. Clyde, however, would prefer to move away from $(t, -s)$. For the initial condition $(-s, t)$ Clyde is the only one who does not wish to move away. Thus neither starting point is a foresighted equilibrium point.

5.4. Gridlock on the Internet

> Picture a pasture open to all. It is to be expected that each herds-man will try to keep as many cattle as possible on the commons. . . . Therein is the tragedy. Each man is locked into a system that compels him to increase his herd without limit—in a world that is limited.
> —**Garrett Hardin**, *The Tragedy of the Commons*

The acidic meadows of the Middle Ages with their herds and their tragedies have been replaced by the binary expanse of the internet. Egotistical academics graze—as did once their oafish forefathers—their electronic herds on the common pasture, to launch them on an equally tragic course.

Blackadder Online (by Alexander Mehlmann)

The sound of ping *beats 'cross the space,*
Good folk, lock up your Unix server,
Beware the deadly interface,
Unless you want to lose your fervor.
Blackadder, Blackadder,
He never needs a guide.
Blackadder, Blackadder,
No Intel chip inside.
He's crawling on the internet
At speeds a million bytes a second.
A virus is his favorite pet
Infecting any site that's fecund.
Blackadder, Blackadder,

> *Doom of the WorldWideWeb.*
> *Blackadder, Blackadder,*
> *You academic pleb.*

The great equalizer on the internet is the daily traffic jam on the information superhighway. Even the considerable increases in bandwidth, for which we can thank the armies of technical innovation nerds, can barely keep pace with the stampede of trampling herds of data. Basically, this is Hardin's well-known tragedy of the commons—a form of prisoner's dilemma for any number of players.

Game theory[6] has already shown a way out of this dramatic impasse. Since the disaster on the internet is based on the fact that no player feels the urge to restrict himself with respect to his internet usage, it seems that the only feasible solution is to introduce something approximating the true cost of internet use.

How much value does a given user x place on the successful transmission of his or her data packet? If he (or perhaps she) wishes to download the latest photos of the Spice Girls during internet rush hour, our user may just have to deal with the failure of that project. If it is a matter of an urgent business deadline that cannot be met because of an overburdened data network, then we shall certainly hear some complaints about lost profits.

MacKie-Mason and Varian [57] demonstrate how gridlock on the internet can in principle be controlled. They suggest that prime-time users become involved in a game that amounts to a sort of specialized auction.[7] Here the strategy of a user consists in making an offer—the price that he would be willing to pay for preferred handling in the case of a data traffic jam—for every electronic cow that he wishes to graze in the internet meadow.

If there is then a data jam at a network node, then a priority queue is formed that is based on the price offered. Only the k highest bidders will be let through, and the price they must pay for service is the bid of the first rejected data packet (that of the $(k+1)$st bidder). This auction rule may seem paradoxical, yet it is extremely effective.

---

[6] In its economic vestments.

[7] Called the Vickrey auction after the winner of the 1996 Nobel Prize in economics.

If it is put into effect, then no user has any reason to offer any bid other than the one that represents the value that he or she has placed on successful transmission of the data packet.

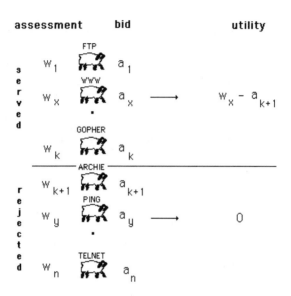

Figure 5.13. Auction on the internet

This is at once apparent from Figure 5.13. If the user's bid a_x is greater than a_{k+1}, the bid of the first rejected internet surfer, then his utility amounts to $w_x - a_{k+1}$. So long as the value w_x does not go below the bid a_{k+1}, he can reach the same utility by offering $a_x = w_x$. However, if $w_x < a_{k+1}$, then every bid a_x that is greater than or equal to a_{k+1}, and therefore also $a_x = w_x$, would have reached the same higher utility 0. Let us now turn our attention to a loser of the internet auction, whose bid a_y is less than or equal to a_{k+1}. So long as w_y does not exceed the bid a_{k+1}, our loser can achieve the same utility by offering $a_y = w_y$. On the other hand, if $w_y > a_{k+1}$, then every bid $a_y > a_{k+1}$ and thus also $a_y = w_y$ would have attained the same higher utility.

The essential mechanism of this auction game is thus the compulsion to offer a bid that corresponds to the value placed on what is to be gained, which can be rationally explained in terms of a weakly dominant mode of play.[8] Furthermore, in this type of auction none of the bidders must fear that he will pay too high a price. Such auctions, also known as "Dutch auctions," were studied and classified in the 1960s by William Vickrey.

William Vickrey

William Vickrey was born in 1914, in Victoria, British Columbia. He studied at Yale and then at Columbia University, in New York, whose faculty he joined in 1946 and where he remained until his retirement. Three days after receiving the news that he had been awarded the 1996 Nobel Prize in economics (jointly with James A. Mirrlees), Vickrey died while en route to a conference.

---

[8]The optimal offer to be made by each bidder proves to be a strategy that **weakly** dominates every other offer in the sense of footnote 7 in Section 2.1.

Chapter 6

Paradoxes of Backward Induction

> He thought he saw a Nash Profile
>   Remaining unrefined:
> He looked again, and found it was
>   Induction from Behind.
> "Before more doubts arise," he said,
>   "Apply it! Never mind!"
> —**Alexander Mehlmann,** *The Mad Reviewer's Song*

A certain concept of solution has achieved a prominent place among the refinements of the Nash equilibrium. Selten's subgame perfection, which in 1965 first saw the light of printer's ink within the covers of *Zeitschrift für die gesamte Staatswissenschaft* [**84**], converted players to plausible modes of behavior even in regions (of the game tree) remote from the straight and narrow path of equilibria.

A tried and true tool for calculating subgame perfect strategies was standing at the ready. Already in 1913 Zermelo [**96**] had proposed methods for the analysis of chess whereby starting from favorable endgame situations one could employ a sort of "backward induction" to arrive at, at least theoretically, the (pure) optimal strategies beginning at the opening move. Backward induction was also the method of choice in Bellman's dynamic programming for ensuring the principle of optimality. Not least, the idea of the Stackelberg equilibrium in a duopoly made economists familiar with this modus operandi.

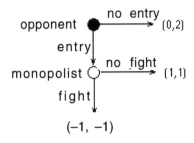

Figure 6.1. Game tree of the market-entry game

The following simple market-entry game will serve us well as a basic component in a complex competitive situation—above all in explaining paradoxical consequences of backward induction.

6.1. The Market-Entry Game

A firm sees its hitherto unchallenged market position (monopoly) threatened by a competitor who has two options: to enter the market or not to enter it. If the opponent attempts to penetrate the monopolist's market, the latter can react either with a strong counterattack or by resignedly putting up with the new market situation. In the latter case the profit that had gone entirely to the monopolist will now be divided fairly between the two participants, while a clash between them would produce detrimental effects on both of them.

As depicted in Figure 6.1, the market-entry game can be formulated as an extensive game with perfect information and memory. On account of the particularly simple sequence of moves, an interpretation as a Stackelberg game immediately suggests itself. In this latter case the potential intruder takes the role of Stackelberg leader, while the monopolist finds him- or herself in the position of Stackelberg follower.

For this reason the strategies of the column player in the normal form representation of this extensive situation will be interpreted as

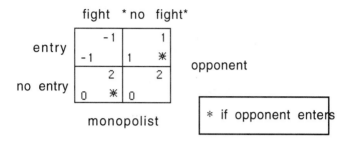

Figure 6.2. Normal form of the market-entry game

reactions to the moves of the row player. Thus for example, let **F** indicate that the monopolist puts up a fight, and **E** that the competitor enters the market. Thus **F/E** denotes the situation that the competitor enters the market and the monopolist reacts by putting up a fight. We will let ¬ stand for the negation of an action, whence, for example, ¬**F/E** represents the case of the monopolist not fighting upon the competitor's entry into the market.

The outcomes marked with an asterisk in Figure 6.2 indicate the equilibria of the bimatrix.

The market-entry game possesses, as is apparent from the normal form, only the two Nash equilibria (¬**E**, **F**) and (**E**, ¬**F**). We recall that in the extensive formulation a strategy is understood as a set of instructions that recommend a unique action to a player for every possible game history that brings him to the current move. Such a recommendation has clearly to be forthcoming even for histories that would never occur by an explicit application of the strategy.

The equilibrium (¬**E**, **F**) illustrates this situation particularly clearly. The strategy of the monopolist recommends to him the action **F** at a decision node of the game tree that is off the path leading to equilibrium.

The equilibrium (¬**E**, **F**) has, moreover, the considerable drawback of eliciting incredulity. It represents the decision on the part of

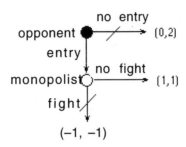

Figure 6.3. Backward induction in the market-entry game

the monopolist to fight to the death, a decision, however, that immediately fizzles out the moment the Stackelberg leader decides to enter the market.

The exclusion of unbelievable equilibria is a trick that can be accomplished through backward induction by following the principles of subgame perfection. By carrying out backward induction for the market-entry game, as shown in Figure 6.3, only the plausible (and simultaneously subgame perfect) equilibrium ($\mathbf{E}$, $^{-}\mathbf{F}$) survives.

However, if the players are given the opportunity, as we have already described in the case of the prisoner's dilemma, to learn through repetition, then we are back at square one vis-à-vis the question of the plausibility of the subgame perfect solution.

6.2. The Chain-Store Paradox

Selten [**86**] proposed the following interesting extension of the market-entry game.

After a successful bout of expansion by which he came into possession of a number of branch stores that together can be considered a local monopoly in various markets, the owner of this collection of chain-stores must again face a well-known challenge.

This time, in each of his branch stores he must deal with a different competitor, where the clash begins with a market-entry game

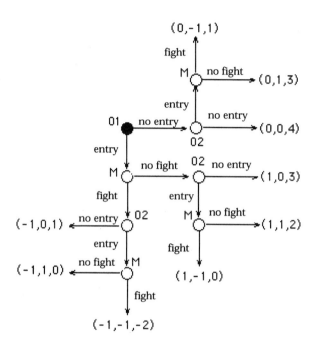

Figure 6.4. Game tree of the chain-store game with two markets

into the first market that is repeated market for market against the next potential intruder.

The game tree of the chain-store game is difficult to keep track of even in the case of a modest number of branch stores. The tree for the case of two stores is described in the following. We have added a new extensive market-entry game at each terminal leaf of the original game tree.

Backward induction yields for the game in Figure 6.4 a unique, subgame perfect Nash equilibrium that suggests to the chain-store owner that he make his peace with the intruder into every market.

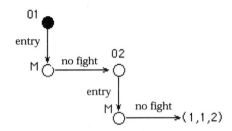

Figure 6.5. Subgame perfect equilibrium path in the chain-store game

The path to equilibrium depicted in Figure 6.5 can be continued in the same manner for arbitrarily (but finitely) many markets. Intuitively, this can be established as follows.

Independent of the history, at the last market an assured entry into the marketplace by a competitor will not be challenged. Namely, a fight would lead only to costs, and since the game has no continuation, this can have no signaling effect for additional markets. This means in effect that the horizon of the chain-store game is shortened by one period, and the arguments that we have made hold without change for the original penultimate marketplace. By logical reasoning we come eventually to the start of the chain-store chain.

The greater the number of disputed markets, the sooner the loss in plausibility of the subgame perfect solution. To understand this effect, we shall step quite a distance off the (subgame perfect) path of equilibrium to consider the case of a chain-store game that is decided sequentially in sixty-six markets.

A potential intruder into the market whose turn it is, for example, at the thirteenth market may have observed the prehistory depicted in Figure 6.6. How should the thirteenth competitor behave? Intrude, as prescribed by backward induction? The monopolist for his part should eschew a fight. But in the first twelve markets this did not happen: The monopolist left one market after another in a shambles without apparently batting an eyelash over his losses.

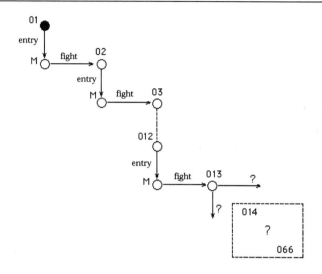

Figure 6.6.   The prehistory of the thirteenth market

If he considers only his reputation, then to be sure, at the thirteenth market the battle cry is "fight!" if only to frighten off the remaining fifty-three competitors. Could this really be only a fight by mistake? A strange error indeed, which he has made with regularity twelve times in a row. Is he mad, then, or worse, irrational? Can a person put his or her trust at all in the rationality assumptions of backward induction?

6.3. The Centipede Game

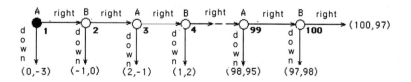

Figure 6.7.   The centipede revisited

We have already seen in the first part of the book an example similar to the one we are about to consider. Are we now with Rosenthal's centipede game in such a completely different situation compared to that of the chain-store game? While the signals that come from a monopolist who neglects the logic of subgame perfection can be interpreted primarily as a means of intimidation, every prehistory in the centipede game that contains a move to the right yields positive impulses for an intuitively anticipated collaboration.

In experiments with this game it has been confirmed in a variety of situations that players are fully amenable—in contrast to the theoretically predicted behavior—to risk a small loss to pave the way for the possibility of a considerable profit. The question of rationality appears again in longer prehistories in which both players have repeatedly acted against the commands of backward induction.

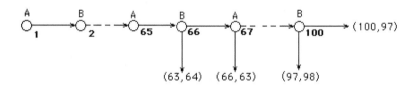

Figure 6.8.   Prehistory in the centipede game

At decision node 66 it would be quite difficult to restore B's faith in the reasonableness of the subgame perfect strategy. Has not player A repeatedly proven that he is prepared to make the move to the right to achieve the higher payoff values that can accrue through trusting mutuality? But is B, on the other hand, prepared to carry out his proven irrationality into the hundredth node? His last move comes like the "amen" in a prayer and showers on the opponent A only 97 utility units instead of the expected one hundred. But wait, has not the principle of backward induction—through the back door, so to speak—entered the playing field?

David Kreps [**51, 52**] has proposed an approach that seems to bridge the deep chasm between the experimentally observable modes of play and the theoretically recommended solution.[1]

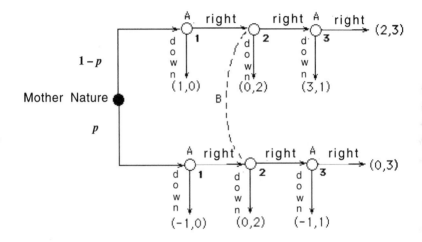

Figure 6.9. The three-pede game with information deficit

In Figure 6.9 we have carried out this proposal for the case of a three-pede. Player B hypothesizes that player A harbors in his breast two souls. One of them is rational and egoistic, while the other is oriented from stem to stern toward cooperation. Both orientations can be simulated by an appropriate utility ordering. Therefore, the first orientation has exactly the same utility values as in the original centipede game. The second, however—because of the assumed utility values—will always surrender to the move to the right. In this way rationality is never at risk for either of the souls.

While player A has a private information advantage at his disposal (he knows which of his two souls will be activated by the vagaries of nature) player B knows only the probability with which one or the other soul will be chosen. The knowledge of this a priori distribution

---

[1]In the game theory literature one observes, however, that in order to achieve success with this approach it would be necessary to give up the assumption of perfect information. Thus this success is achieved at all too high a cost.

is common to both players; we will deal with the properties of this game-theoretic notion in Section 7.2.

On the basis of the subgame perfection the strategic behavior of both souls is foreseeable at the third decision node. The egoistic soul will move downward, while the cooperative soul will choose the rightward move. The cooperative soul will also prefer the move to the right in the first node, since it thereby with certainty—what B will always undertake in the second node—can only win. The result of this combination of backward and forward induction is depicted in Figure 6.10.

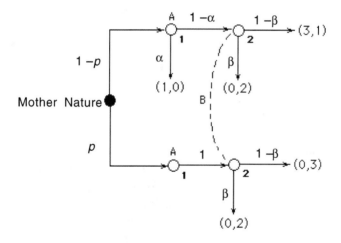

Figure 6.10. The reduced three-pede game

An equilibrium in behavioral strategies can be immediately deduced from the game tree shown in Figure 6.10. Namely, if the egoistic soul at node 1 is indifferent between the move downward and the move to the right, then the utility equation $1 = 3(1 - \beta)$ should hold. Thus at equilibrium, B chooses the move to the right with probability $\frac{1}{3}$. On the other hand, the indifference of player B can also be put to good use.

According to his conjecture, he finds himself with the a posteriori probability

$$\gamma = \frac{p}{p + (1 - p)(1 - \alpha)}$$

in the lower node of his information set. Since he should obtain the same utility independently of his choice of move, the equation $3\gamma + 1 - \gamma = 2$ holds. At equilibrium A's egoistic soul thus chooses the move to the right with probability $p/(1 - p)$ (for $p < \frac{1}{2}$).

One should note that even the tiniest doubt with respect to the intentions of the opponent suffices to induce both players to deviate with positive probability from the path of equilibrium. The same effect can be observed in the chain-store game. This is not in the least surprising, since in both cases players who are repeatedly offered new decisions can carefully build up their reputations.

Paradoxes of backward induction can nonetheless also appear in conflict situations where players cannot observe the whole history of the game. In this class of games the level of future expectation blurs every influence of the past game development. The story of the bottle imp analyzed below proves to be a meaningful parable for such mechanisms.

6.4. The Bottle Imp Paradox

> The words died upon Keawe's tongue; he who bought it could never sell it again, the bottle and the bottle imp must abide with him until he died, and when he died must carry him to the red end of hell.
> —**Robert Louis Stevenson**, *The Bottle Imp*

Once upon a time, there was a man who dwelt on the island of Hawaii. One day, he was offered for purchase a bottle that was said to possess wondrous properties. The devil himself was said to have set it on its course in the world, and in former times it was acquired at an enormous price. However, though it was now being offered at a bargain-basement price, its value should really be set much higher, for within the bottle's interior dwelt the devil himself (see Figure 6.11), who would grant whosoever might own the bottle any wish that he or she might care to make. But should the owner die before he can rid himself of the bottle by selling it for minted coin at a price less than what he paid for it, then he will be cast directly into hell.

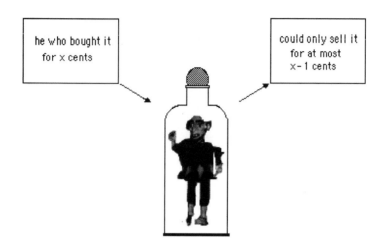

Figure 6.11. The price of damnation

Keawe—for that was the man's name—hesitated at first. But since the price of the bottle had sunk over the centuries to a mere $89.99 and since he figured that should the need arise he would have a good chance of finding a buyer,[2] Keawe and the bottle's owner finally closed the deal.

A game-theoretic angelus ex machina might at this juncture have explained to Keawe that his decision ran counter to the principle of backward induction. If the bottle were to be offered at a price of one cent, then it is clear that a buyer would not be found, since there is no coin of lesser value in circulation.[3]

---

[2]Even though in good conscience he would feel himself duty bound to explain the disadvantages that might accrue to the owner of the bottle.

[3]In Stevenson's original, Keawe, after successfully making use of the bottle and selling it and later falling ill with leprosy, in fact acquires the bottle for the price of one cent. His disease is cured by the magic of the bottle. Keawe then falls into deepest despair over the prospect of eternal damnation. Then his wife remembers that in Tahiti one can exchange one cent for five French centimes, and thus the bottle-selling game finds its dramatic continuation.

Yet if no one is prepared to purchase the bottle for one cent,[4] then surely no one would purchase it for two cents, and inductively, no one should be willing to buy it for any price at all.[5]

The crucial difference between the centipede game and the bottle imp game lies in the rule that is supposed to induce subgame perfect behavior. In the bottle imp game the subgame perfect strategy *do not purchase* does not stand in opposition to the previous development of the game, since the player in question has not been exposed to multiple decisions as to whether to purchase. On the other hand, in the centipede game player A, for example, will be obligated to move downward at the ninety-ninth decision node, although he knows perfectly well that he—because of his own subgame perfect good behavior at the root of the game—will never be in a position to execute this move.

Now, classical game theory attempts (paradoxically) to drive out the paradoxes of backward induction precisely with the aid of such decisions against the course of events. In Section 7.3 we shall surrender under resolute protest to these intricate lines of argumentation.

6.5. Rituals of Division

> Let them eat cake.
> —**Marie Antoinette**

> "I say, this isn't fair!" cried the Unicorn, as Alice sat with the knife in her hand, very much puzzled how to begin. "The Monster has given the Lion twice as much as me!" "She's kept none for herself, anyhow," said the Lion.
> —**Lewis Carroll**, *Through the Looking Glass*

---

[4] In the end, Stevenson brilliantly breaks through the backward induction. Since no one was willing to purchase the bottle for four centimes, Keawe's wife purchases it through a front man. Keawe, who later sees through this masterstroke, commissions a helmsman to get his wife off the hook for two centimes with the understanding that he will then purchase it for one centime. This front man then resolutely refuses to part with the bottle, since the prospect of the journey to hell does not frighten him in the least (an irrational response?).

[5] According to this logical argumentation no individual should be willing to enter into a pyramid scheme or—what would be considerably more advantageous—to launch one. Can one therefore describe the existence of well-heeled con men and (much more numerous) shorn sacrificial lambs as a paradox? In the special case of a pyramid scheme certainly not, for the instigators of such schemes are determined in the end to grant their participants only an incomplete view of the scheme's structure.

The most famous problem of division in the Bible has as its theme more the wisdom of the divider than the question of fairness. The judgment of Solomon can thus be seen as a worthy precursor of signaling games. The wise advice to slice the indivisible baby into two equal portions was intended above all to have the following effect: to reveal which of the two women was the true mother and which the impostor.

Without doubt the signal was clear, understandable, and foreseeable to the true mother.[6]

However, King Solomon's easy success in his game with the two mothers can be attributed above all to the fact that the false mother reacted naively, shortsightedly, and in a manner in extreme conformance with the demands of the parable.[7]

How would the wisdom of Solomon have fared if each mother—in genuine or feigned concern for the life of the infant—had surrendered her claims to the child in favor of the other? A more or less wise judge in our era would probably have given each woman limited visiting rights to the child and entrusted the infant to the care of the state.

Instead of threatening to halve the infant, a despotic monarch would also bring into play the threat to draw and quarter the two mothers. Glazer and Ma [**34**], on the other hand, have drawn up a simple extensive game for a game-theoretically savvy Solomon, a game that achieves the goal of weeding out imposture without any inquisitorial interrogations under threat of torture but only under the threat of exposure to some degree of pecuniary sacrifice.

In Figure 6.12 Solomon asks the first woman whether the infant is hers. If the answer is no, then the child is awarded to the second woman. The second woman's (from Solomon's point of view unknown) utility $W_{??}$ amounts to W_t units if she is the true mother and W_f units if she is the false one. In this case the first woman achieves a utility of only 0. On the other hand, if she replies, "yes, it is mine," then the same question is put to the second woman.

---

[6]Thus spake the true mother: "O my lord, give her the living child, and in no wise slay it."

[7]Thus spake the false mother: "Let it be neither mine nor thine, but divide it."

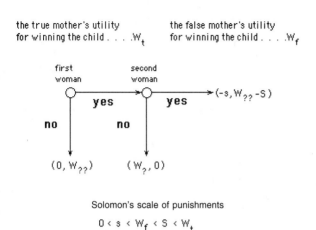

the true mother's utility
for winning the child . . . .W_t

the false mother's utility
for winning the child . . . .W_f

Solomon's scale of punishments

$$0 < s < W_f < S < W_t$$

Figure 6.12. The judgment of Solomon after extensive questioning

If she answers no, then the child is awarded to the first woman, whereby her utility $W_?$ again amounts to either W_t or W_f. Clearly, the second woman goes away empty-handed.

However, a second "yes" answer yields interesting consequences. Instead of a child, the first woman gets a punishment p, which is nonetheless smaller than the utility value of a successful false mother. Likewise, the second woman is punished, though she has won the child. Her punishment P is set at an amount between the utility values W_f and W_t.

These punishments, known in advance, have the effect of a truth serum. Namely, if the first mother is the false one, then she knows that the true mother will answer yes at the second decision node (since $W_{??}(= W_t) - S > 0$). Thus in the first node she must confess the truth, namely "no," since she is under the threat of a loss of p utility units.

In a scenario in which the roles are reversed, the second mother will avoid at all costs the answer yes (since $W_{??}(= W_f) - S < 0$). In expectation of this, the first mother will stick to the truth and assure herself of the maximum payoff $W_? = W_t > 0$.

The fully explainable suitability of the mechanisms of backward induction in contributing to the establishment of truth in such a complicated situation should nevertheless not blind us to the fact that one is also at the mercy of these mechanisms' paradoxes in the environment of problems of division.

A classic example of this—propagated in the literature [**35**] under the foreboding appellation "ultimatum game"—has captured the attention of game theorists. Two players attempt to agree on a mutually satisfactory division of a cake.[8] The first player offers the second a portion of size x. If this portion is accepted by the second player, then the cake will be divided accordingly. Otherwise, both come away empty-handed.

If we now assume that the cake can be divided into at most a finite number m of crumbs of size k, so that the possible offers by the first player can be represented as $x = i \cdot k$, $i = 0, 1, \ldots, m$, then the process of backward induction yields two subgame perfect equilibria for the extensive game in Figure 6.13.

For every offer, starting with a single crumb, the best answer of the second player is to accept the offer. On the other hand, if nothing is offered, then the player is indifferent as to accepting or rejecting. In the first subgame perfect equilibrium the second player is offered nothing—not even a crumb—an offer that is not rejected[9]

---

[8]In the simplest case of a procedure by which two persons fairly divide a cake between themselves, the first player is permitted to divide the cake in two, while the second is then granted the right to choose one of the two portions. If the number of hungry mouths exceeds two, then the division procedure becomes considerably more complicated. See Hugo Steinhaus's [**89**] suggestions for a troika of sweet tooths (sweet teeth?) as well as (for the most general case) the last, sweetest revelation, the recipes of Steven Brams and his slice-happy pastry sous-chef Alan Taylor [**13, 14, 15**]. Before carrying out the given instructions at a birthday party, say, among a company of considerable size, beware: At the first step a cake intended to satisfy the cravings of n hungry partygoers must be divided into $2^{(n-2)} + 1$ pieces. Since the number of subsequent divisions necessary for an eventual distribution of what will amount to a rather large number of crumbs is extremely large, it will take so long for the distribution of the cake that the host may be faced with a moderate-sized rebellion, or else possibly will have managed to make the remains so unattractive as to spoil the appetites of all of his guests, thereby simultaneously solving the division problem and offering a secular explanation of the miracle of the loaves and fishes.

[9]This comes as no surprise, since in the second player's calculations of his utility there is no place for sticking it to the other player.

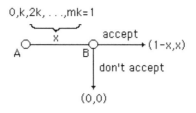

A ... first player

B ... second player

Figure 6.13. The ultimatum game

(and this is thus the best response of the first player to the expected acceptance). In the second equilibrium, on the other hand, the first player calculates that the offer of nothing will be rejected (which would represent the loss of an entire cake) and selflessly gives up a crumb of size k. The second player accepts this generous offer (in the silent hope, perhaps, that the first player will choke on the first of his $(m-1) \cdot k$ crumbs.

The outcomes of experimental studies of the ultimatum game stand in stark contrast to the results predicted by backward induction. In [**78**] there has even been established a regional differential in the readiness of the first player to offer the second more than is recommended by the theory. The interpretations of these results seem to point in the direction of evolutionary behavioral norms that are the expression of heterogeneous cultural and social characteristics. We have now to fear that the centrifugal force of globalization will eliminate this diversity of behaviors in favor of a myopic rationality.

Chapter 7

Strategic Accents of Game-Theoretic Scholasticism

> He thought he saw a Strategy
> Undominated, strict:
> He looked again, and found it was
> Quite Easy to Depict.
> "I'll never play a game," he said,
> "So simple to predict!"
> —**Alexander Mehlmann,** *The Mad Reviewer's Song*

The extent to which the strategic accents of game theory become a mixed question of belief will be demonstrated in this chapter with the help of various sources of game-theoretic scholasticism.

7.1. Seeing Through the Opponent

> **Blackadder:** It's the same plan that we used last time, and the seventeen times before that.
> **Melchett:** E-E-Exactly! And that is what's so brilliant about it! We will catch the watchful Hun totally off guard! Doing precisely what we have done eighteen times before is exactly the last thing they'll expect us to do this time!
> —**Richard Curtis and Ben Elton,** *Blackadder Goes Forth: Captain Cook*

The ability to look ahead or anticipate has already been described in all its ambiguity in Conan Doyle's *The Final Problem* [**18**].

Sherlock Holmes and Professor Moriarty confront each other in a sort of prelude and candidly discuss their common knowledge.

It is for Moriarty, the mathematician, to propose the first conjecture: "All that I have to say has already crossed your mind." To which the world's first private consulting detective answers drily, "Then possibly my answer has crossed yours."

Here intellect opposes intellect. Indeed, even the subsequent pursuit to Dover degenerates for all its dynamic drama into a masterly duel of anticipation. Holmes, whose life hangs in the balance, flees in an express train to Dover in order to reach the security of the Continent. Moriarty divines that Holmes is leaving from Victoria Station. Holmes is convinced that his foe has seen through him and will follow him to Dover in a private train.

In order to evade this move of Moriarty's, Holmes alights at Canterbury. At this point, alas, Conan Doyle alights from the carousel of anticipation in order to bring his tale to a conclusion. In order to save the good names of Moriarty and Holmes, Oskar Morgenstern, one of the fathers of game theory, continued the game of seeing through the opponent in the following manner.

If Holmes has decided to alight in Canterbury, then Moriarty should "again do what I would do" and also alight at that station. Anticipating this, Holmes should continue on to Dover, which should induce Moriarty not to get out at Canterbury, which should give Holmes the idea to disembark at that stop after all, and so on, and so on, and so on ... [1]

Thus Morgenstern arrived at the naive conclusion that there is no way out of this vicious circle of mutual anticipation. Meanwhile, however, we know better (see Section 1.1), and we ultimately have Borel [11] and von Neumann and Morgenstern [71] to thank for this knowledge. To unravel the Gordian knot of decision-making in the context of perfect anticipation we require the Alexandrian sword of chance.

Thus can the anxiety of the goalkeeper before a penalty kick ultimately be described as the fundamental fear of making the wrong

---

[1](... and so on.)

decision in the cycle of anticipation.[2] The Chinese goalkeeper Gao Hong should lastly throw herself into the left corner of the goal to deflect Brandi Chastain's (United States) decisive penalty kick on the assumption that perhaps the female former forward tended to kick to the right.

On the assumption that soccer players are rational beings and moreover are inclined to plan ahead, Brandi could decide on the other corner or slam the ball down the middle of the goal. A profound analysis of further anticipation possibilities can be left at this point to much more knowledgeable sports commentators.

While we may understand the effort to see through the opponent as an important component of a strategy, the introduction of chance leads to contradictory interpretations of strategic modes of thought.

From Aumann [2] we have an illuminating explanation for the reasonableness of mixed strategies of normal form. The mixed strategy associated to one player will then be seen not only as a random selection from among his pure actions, but above all as a *belief* shared by all others of his patterns of behavior. Thus in a mixed Nash equilibrium every action (chosen with strictly positive probability) of a player is the best reply to his own beliefs about the behavior of the opponent.

In accord with this picture, games are above all decided in one's own head. An example that is unusual in many respects for this idiosyncratic form of game-theoretic deduction can be discovered in Gregor von Rezzori's Maghrebinian stories [77].[3]

One day, the rabbi of Sadagura is unexpectedly called into town. He thus finds himself confronted by the following conflict situation. The meager portion of meat that he has set aside for his midday meal will probably in his absence attract the attention of his dog, Bello.

---

[2]In [25] quoted from Peter Handke, *Die Angst des Tormanns beim Elfmeter.*

[3]The extent to which this unfortunately greatly underappreciated chronicler disguised the truth can be best judged by the author of these lines. When the Maghrebinian robber Terente (who carried out his infamous deeds not only in the Maghrebinian stories but in the real world as well) was caught in a police trap, my grandfather, at the time bearing the title of Imperial and Royal physician, was entrusted with his care. With tears pouring down his cheeks the Maghrebinian Robin Hood begged for medical treatment and underscored his request with the observation that he had been a classmate of my father's. *Non scholae, sed vitae discimus.*

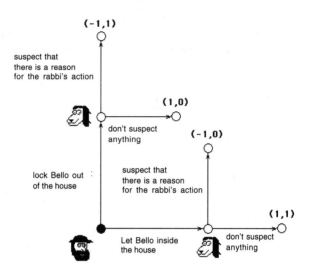

Figure 7.1. Seeing through the opponent—Maghrebinian variant

The only options that are available to him are these: (1) Lock Bello out of the house; (2) let him remain inside. If the rabbi chooses the former, then (in his opinion) Bello will at once conjecture that there is a reason for this unusual action. Therefore, he will strive to get into the house, and he will thereupon sniff out the meat and eventually devour it. If the rabbi chooses the second option, then Bello will have no reason to suspect anything and will thus neither sniff out the meat nor eat it.

In Figure 7.1 we have presented the game as it should proceed from the point of view of the wise rabbi. The solution would be to leave Bello in the house, which, so I have heard, is what indeed happened. When the rabbi returned home, Bello had eaten the meat. Then the rabbi turned to his dog, tapped him on the forehead, and said reproachfully, but gently, "Bello, you have your head screwed on backwards."[4]

7.2. Common Knowledge

> He thought he saw a Dirty Face
>> In common knowledge rage:
> He looked again, and found it was
>> A Game Without a Sage.
> "Just read this book," he faintly said,
>> While blushing at each page!
> —**Alexander Mehlmann,** *The Mad Reviewer's Song*

Before the game began, the rabbi of Sadagura was (all too) certain that Bello was a rational player in the extensive game of Figure 7.1. This certainty, however, seems to have been somewhat one-sided. To play out equilibria in extensive or normal-form games one requires certainty of a higher quality.

In their tireless search for the grail of an interactive epistemology, Aumann and Brandenburger [4] have isolated the following sufficient conditions for the emergence of an equilibrium for the case of a two-person game in normal form: Every player must be certain of the utility values, the impressions, and the rationality of his opponent. This cognitive level will also be called *mutual knowledge.*

However, as soon as the number of individuals involved in a game exceeds the number two, a state of mutual knowledge will no longer get you very far. To approach an equilibrium, not only must a player know particular characteristics of his opponents; the opponents must be certain of the first player's certainty as well; he, in turn, must be certain of his opponents' certainty; and so on, ad infinitum.

Long before the notion of common knowledge captured the imagination of the decision theorists and game theorists, it gave the world something to ponder in the form of logico-mathematical puzzles. The oldest example comes down to us from Littlewood's collection [**56**]. It is a tale of three (Victorian, perhaps?) ladies who are seated in a compartment of a passenger train and suddenly break into hysterical laughter, each on account of the soot-stained faces of the other two. Suddenly, the laughter of one of the ladies freezes in her throat,

---

[4] A doubt conceived in hindsight as to Bello's rationality?

and she blushes for shame. We shall explain later why this strange behavior is an immediate consequence of common knowledge.

First, we would like to interpret a modern variant of this story.

The Kakanian Mole

The fifth column of the Kakanian[5] secret service, in comparison to which even the French Army's Deuxième Bureau can be considered only second class, had in its heyday more moles than London's MI-5.

To bring an end to this untenable state of affairs, a specialist in counterespionage was smuggled into the column. In less than a week he summoned his associates and gave the following report:

"Meine Herren! At least one mole has been discovered among us. He himself does not suspect this, but the name of each exposed mole has been provided to all his colleagues. As soon as the individual or individuals in question have attained certainty that they have been discovered, they will have precisely one hour to draw the necessary conclusion. Go each of you into your own office and leave it only when you hear the report of a service revolver."

Thirteen hours later shots ring out from the offices of the fifth column. What has happened?

To solve the Kakanian riddle we must evaluate the informational situation of the individuals in the fifth column. Before the counterespionage specialist made his announcement, every employee had been informed as to which of his colleagues were exposed moles. However, before the announcement, this information was one-sided certainty. Only with the announcement was the process of recognition set in motion. The information that there was at least one exposed mole became common knowledge.

Now, the number of exposed moles was certainly greater than one. For otherwise, the lone mole who had no information about one

---

[5]Translator's note: The *kaiserlich und königlich* "dual monarchy" of Austria–Hungary was an arrangement arrived at by way of the *Ausgleich* (compromise) of 1867 under which the Hapsburg emperor (hence *kaiserlich*, i.e., imperial) ruled over the twin kingdoms (hence *königlich*, i.e., royal) of Austria and Hungary. The universal abbreviation "K u. K" evidently encouraged Robert Musil, in *The Man Without Qualities*, "Ka" being the pronunciation of the German letter "K," to coin the mildly scatologic-sounding nickname "Kakania."

or more other exposed colleagues would have known at once that he had been exposed and would have shot himself at the end of one hour.

Had there been only two exposed moles, then everyone would have received the names of the two exposed colleagues, except for the two exposed moles, who would have received one name each. However, since the number of exposed moles was at the outset a matter of common knowledge for no one, the first hour passed without a sound. At the beginning of the second hour each of the two who had been given exactly one name would have attained certainty that he had been exposed. At the end of this hour there would therefore have been two shots.

If we assume that $k - 1$ exposed moles will shoot themselves at the end of the $(k-1)$st hour, then the k employees of the fifth column who were given only $k - 1$ names would conclude at the end of the $(k - 1)$st hour that they had been exposed. During the course of the next hour they would load their revolvers, set down their last will and testament, and then go to meet their makers.[6]

Thus, by means of complete induction, has the riddle of the Kakanian mole been solved. There were exactly thirteen exposed moles.[7]

The function of the public occasion that puts everyone in a position to attain common knowledge is replaced in Littlewood's puzzle by the women's hysterical laughter. Thereby it is clear to all three travelers that at least one of them has a sooty face, and that all know that they know it, and so on. Without a logical sequence of blushing, which we clarify in Figure 7.2, Littlewood's solution of the puzzle has no basis.

If situation b, c, or d obtains, then one and only one woman has a sooty face. The other two (and only these two) have a reason to laugh. Thus the woman with the sooty face should blush at once, since it is apparent to her that the others are laughing at her.

---

[6]In fulfillment of the strict code of honor to which Kakanian moles were subject.

[7]Those who might find the example of the Kakanian mole somewhat macabre must be unfamiliar with the more classical variants. In [65] forty women shoot their philandering husbands after they have attained common knowledge that in their matriarchal kingdom there is at least one faithless husband. In [33] infidelity on the distant planet Womensa is punished by castration and public exhibition.

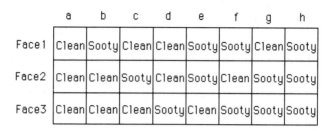

	a	b	c	d	e	f	g	h
Face1	Clean	Sooty	Clean	Clean	Sooty	Sooty	Clean	Sooty
Face2	Clean	Clean	Sooty	Clean	Sooty	Clean	Sooty	Sooty
Face3	Clean	Clean	Clean	Sooty	Clean	Sooty	Sooty	Sooty

Figure 7.2. The state of affairs in Littlewood's puzzle

On the other hand, if we find ourselves in situation e, f, or g, then the two soot-faced women each wait to see whether the other will blush. Each at first conjectures that situation b, c, or d might be in force (and that she herself has a clean face). But since the other woman does not blush, this initial conjecture is seen to be false (and one has oneself a sooty face). At that moment of recognition, the two women blush simultaneously.

In situation h all three of the ladies wait to see whether the other two blush simultaneously. When this twofold blush fails to materialize, the certainty finally arrives to each of them that her face is a sooty one as well, and the three of them tranquilly blush in triplicate. Thus Littlewood greatly underestimated the number of blushing ladies.

With common knowledge we have finally described the greater part of the dogmas that hold together the world of games. Generally, certain games[8] make use of a wise person to set all the players on the course of common knowledge. Yet the true wise persons in the game that is game theory are those who can recite the incantations of common knowledge without tying themselves in into logical pretzels. If these axioms are translated into common parlance, they take on the aroma of the litany of a charlatan.

Dogmata of common knowledge

1. *If one knows everything, then there is nothing that one does not know.*

---

[8]the so-called *games with a sage.*

2. *One can know only that which has taken place.*

3. *Before one can know something, one must know that one knows it.*

4. *If one does not know that something did not take place, then one knows at least this.*

If the state of common knowledge is not reached with respect to all relevant informational matters, then consequences of considerable gravity can ensue. A game-theoretic representation of these already known connections from the theory of distributed systems—as a problem of coordinated attack[9]—was taken up by Rubinstein in [**80**].

7.3. Going Against the Course of Events

> He thought he saw an Argument
>   That proved he was the Pope:
> He looked again, and found it was
>   A Bar of Mottled Soap.
> "A fact so dread," he faintly said,
>   "Extinguishes all hope!"
> —**Lewis Carroll,** *The Mad Gardener's Song*

It is certainly no accident that for this section's epigraph we have chosen one of the immortal stanzas of the mathematician Charles Lutwidge Dodgson (alias Lewis Carroll).[10] If he could arise from the dead, this famous writer known for his eccentric logic would assuredly enjoy the discourse about decisions that go against the course of events.

---

[9]This problem is known to have cost Napoleon the victory at Waterloo. The marshals Ney and de Grouchy were unable to coordinate their attacks on the also separately acting armies of Wellington and Blücher. De Grouchy is thought to have sent the following piece of e-mail to ney.marechal@waterloo.mil. *Your Excellency! You deal with Wellington. I'll handle the Prussian. Please confirm.* Upon which Ney sent to de.grouchy.marechal@waterloo.mil the following: *Dear de Grouchy! Attack as suggested. Confirm receipt.* As long as each message had the risk of not being received, the sender had to wait for confirmation, which he in turn must confirm. The coordination of an attack can never achieve common knowledge.

[10]We offer thanks here to Dimand and Dimand [**22**] for the recognition that Lewis Carroll (under his given name) was a significant—even if thoroughly underappreciated—forerunner of game theory. In letters to his political contemporaries and in a refreshing pamphlet on the principle of parliamentary representation, Dodgson revealed himself to be a stylistically and mathematically brilliant thinker who seemed at home with game-theoretic modes of thought sixty years before the discipline appeared.

Unfortunately, in the last sentence we have already made use of a *counterfactual*—a conditional decision that goes against the course of events. The assertion that Carroll would assuredly enjoy the discourse is conditional on an occurrence about which we know that it cannot occur.[11]

Let us for a moment set aside both Lewis Carroll and the question of resurrection in order that we might for the last time make our way through the depths of game-theoretic interpretations. Along the way we shall be particularly interested in the after-effects of the paradoxes of backward induction.

To act in accordance with the strategic rules of backward induction would be in no way a paradox if one could pretend that this mode of behavior was the immediate consequence of rationality. In [3] Aumann plays the strongest card in this regard: the common knowledge that all players are rational.

Now, this common knowledge, formally disguised as a sort of dark force—comparable with the weavings of the three fates—stands behind the actions of the player. If it has once appeared in that timeless enchanted realm that the theoreticians generally call the pregame, then it weighs in its fatal immutability upon the decision-making power of those involved. They do not stray—thus Aumann—from the current path of backward induction even if they (or their fellow players) have already left such a valid path in the past.

What primarily bothers Ken Binmore [9] about this form of argumentation is the dogma of rationality that even through an unambiguous incorrect action of an oh, so rational player is not to be shaken. For him the fall from grace lies above all in the application of formal models. He thus rejects the virtuosity even of someone like Dov Samet, who is an admirable master as much of the entire range of formalism [81, 83] as of the rapier of the pamphlet [82].

With Samet's pamphlet the circle is closed, and we find ourselves again in the company of Lewis Carroll. In the sixth chapter of *"Through the Looking Glass and What Alice Found There,"* we meet

---

[11]If this were an esoteric treatise, then the sentence in question would obviously not be a *counterfactual*. (Oops! Another counterfactual.)

hubris in ovoid form—Humpty Dumpty. Balanced in unstable equilibrium on the edge of a narrow wall, Humpty Dumpty gets carried away by his belief in counterfactuals.

Were it to happen that Mr. Dumpty should fall—quite an improbable event from either the Humptian or Dumptian viewpoint—but just in case he should nonetheless happen to fall, then surely the king's horses and men—according to the Humptian and Dumptian belief system—would be able to put Humpty back in his place again.

All the kings horses and all the kings men, the angels of the Lord, the lord of the flies, the voice of the people, the voice of the turtle, and not least the shared certainty of rationality—all of these should rescue our chestnuts from the fire, which in any event we should never have let fall in there in the first place. But can we truly place our faith in rationality?

Questions like these shake the very foundations of our civilization. Whereas in *Finnegans Wake*[12] Humpty Dumpty is pressed into service as a parable for Lucifer's fall into hell, Samet adds him as a myth of game theory to the great resurrection parables of humankind.

In our fast-paced era the gods (of politics) care more about their resurrection after the current legislative session. Thus arise modern myths of secular interpretation that can be adapted to the fall of Humpty Dumpty.

All of this fades in the face of the primal force of the ancient myths. In the following section we shall analyze the game-theoretic case of the conscientious objector Odysseus.

---

[12] We make reference to James Joyce's cryptic masterpiece without casting any aspersions on the ballad with the same name about the fall of the hard-drinking Irishman Tim Finnegan.

Chapter 8

Odysseus Goes to War

Agamemnon et Menelaus Atrei filii cum ad Troiam oppugnandam coniuratos duces ducerent in insulam Ithacam ad Ulixem Laertis filium venerunt. cui erat responsum, si ad Troiam isset post vicesimum annum solum sociis perditis egentem domum rediturum. itaque cum sciret ad se oratores venturos insaniam simulans pileum sumpsit et equum cum bove iunxit ad aratrum. Quem Palamedes ut vidit sensit simulare atque Telemachum filium eius cunis sublatum aratro ei[us] subiecit. . . .
—**Hyginus**, *Mythographus*, Fabula XCV

"The function of myth," writes Mircea Eliade [**26**], "consists in revealing models and thereby providing meaning to the world and human existence." In eras in which the power of myth has attenuated, the muses of literature and science are summoned to carry out this function. Yet this does not relieve us of the duty to pose the question of the usefulness of a model that analyzes archaic conflicts of a rather insignificant prelude to the Trojan War.

Already in the fifth century B.C.E. the myth of the war of the Achaeans against Ilion had been reduced by Thucydides [**91**] to a mundane question of microeconomics.

If we thoughtlessly follow the version of that battle-tested historian, then hidden behind the epic clash is none other than the optimal deployment of superior capital reserves in the case of a successful attempt at penetration into the western Anatolian slave and ceramics markets.

This golden pons asinorum between economics and the ideal of classical culture has never been surpassed by a guru of game theory. So it is not at all surprising that we comfortably installed academic purveyors of horsefeathers have—instead of a Trojan horse—nothing better to offer than *Selten's horse.*

On the other hand, what meaning can be ascribed to our offering? Methodically, the latter stage of the two-stage signaling game defined in Figure 8.1 exhibits the essential features of the well-known *quiche and beer* model [**16**] without making use of its savory café charm. However, in opposition to In-Koo Cho and Kreps [**16**] we shall not in our work pursue purely didactic goals. The starting point of our considerations is the tale of the madness of Odysseus as described in the ninety-fifth fable of Hyginus [**43**].

8.1. The Madness of Odysseus

On a chilly autumn day in the year 1260 B.C.E powerful oar-strokes steered a Mycenaean battleship into the deserted harbor of Phorkys on the island of Ithaca. Caparisoned in bronze and leather, the high king Agamemnon bestrode the isle. At his side trod Palamedes, the keen-witted son of Nauplios, whom the Achaeans rightly regard as the inventor of the alphabet, dice, and board games.

The harbor authorities were nowhere to be seen. Before the un-manned wooden sentry house there was to behold naught but graz-ing sheep. The son of Atreus was assuredly unprepared for such a reception. In quest of counsel, he glanced over to his beweaponed companions, in response to which glance one of them—a friend fair of speech—loosed the reins that held his tongue.

> "This isle," commenced Palamedes,
>   dactylic hexameters voicing,
> "Is Ithaca's rocky dominion,
>   the home of the hero Odysseus.
> What luck to depart from this landscape,
>   to leave the impoverished hillsides,
> In Ilion's walls to accumulate
>   glory's bright laurels eternal.

The vow that all suitors did swear to,
    shall now be acknowledged in honor,
Our duty to liberate Helen,
    by means of mandated conscription."

Despite all the love that the Achaeans—at least in their epic poetry—have shown for the craft of versification, the situation here is too perilous for Agamemnon (and you, gentle reader) to put up with any more of the author's dactylic doggerel.[1]

Paris, as we know, had carried off the beautiful Helen to Troy. This audacious appropriation of the property of another was displeasing to Helen's cornute husband, and since the suitors of Helen had once taken a solemn oath that they would stand by that husband should his wife one day come into dispute, within a short time a considerable force had been assembled to engage in a campaign of revenge and recovery.

Only a few seemed not to hear the call to arms. One of these, Odysseus, had in fact left three urgent messages from the Achaean general staff unanswered. According to rumor, the Delphic Oracle had prophesied that should Odysseus participate in the war, he would be condemned to twenty years in exile. Now Agamemnon, his designated commander, had arrived in Ithaca to apprise, in person, him who was liable for military service of the general mobilization.

The Mycenaean delegation found the island in a desolate state. An unusually fine vintage was rotting on the vines. The royal palace on the mountain Aetos sheltered only its servants. Descending along the western slope the armed troops were met by Penelope, wife of Odysseus, distraught, in tears, and bearing her infant son Telemachus in her arms.

"Where is your husband, woman?" barked Agamemnon. The queen of Ithaca indicated with a nod of her head the path to the lonely beach below, where a powerful form was plowing meandering furrows into the soft sand. Horse and ox were yoked to the plow. The man who was plowing wore a peaked cap and without surcease was

---

[1]Translator's note: Rendered into English doggerel by the translator.

sowing salt. It was Odysseus, whom the gods had apparently struck with madness.

"Unfit for military service," observed Agamemnon unhappily. Then Palamedes grasped the infant and placed him in front of the plowshare. Would Odysseus run the plow over his son?

8.2. The Game-Theoretic Model

Without wishing to anticipate the further melodramatic events, it is now time to introduce some game-theoretic considerations. We shall interpret Odysseus's dilemma as the result of extensive moves in a two-stage, three-person game with incomplete information. The Mycenaean team will be represented by Palamedes. His opponents are Odysseus the malingerer and Odysseus the madman. At the start of the game Palamedes is lacking the information as to which of these two personae is the true Odysseus.

However, this lack can be compensated by playing a conjurer's trick. In the tradition of Harsanyi [**36, 37, 38**], a random move will be introduced at the beginning of the game.[2] Thus, by throwing the dice according to an a priori distribution known to all, Palamedes' opponent will be chosen.

The information structure of this new game is now complete. However, it is imperfect, since Palamedes knows only the probability p (respectively $1 - p$) that it is Odysseus the madman's (respectively Odysseus the malingerer's) turn. However, his two adversaries always know precisely whose turn it is.

Once fate has made its choice, then Odysseus has the opportunity to make his first signal. Independent of his randomly determined ego type, he can either abstain from any utterance whatsoever—strategy $\bar{m}$—or, if he so chooses, play the madman—strategy m.

Now it is Palamedes' turn. If nothing is signaled to him, then he must decide between renunciation and conscription—strategies r and c. On the other hand, if he receives the signal "madman," then he

---

[2]Harsanyi speaks in general of a move by nature. We, however, should doubtless in the present case be speaking of a move initiated by the Moerae—the three fates: Lachesis, the distributor, who apportions human destiny; Clotho, the spinner, who spins the thread of life; and Atropos, the inflexible, who cuts it.

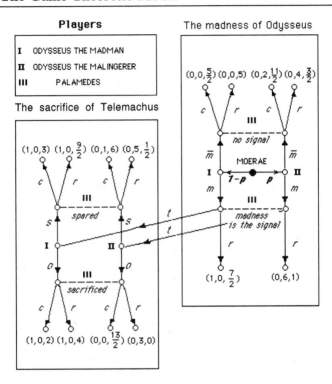

Figure 8.1. Odysseus goes to war: game tree of the three-person game

can either renounce Odysseus—strategy r—or bring Telemachus into play—strategy t. In this latter case, the game takes a dramatic turn. Odysseus is forced to give a new signal. He can offer up Telemachus—strategy o—or spare him—strategy s. Palamedes will then react either with renunciation or with conscription.

Figure 8.1 describes the extensive game by presenting the corresponding game tree. The utility values of the three players indicate their preferences from among the possible game outcomes.

So, for example, Odysseus the malingerer values successfully deceiving his opponent the highest, provided that Palamedes renounces playing the Telemachus card. The second-highest valuation is placed on successful deception followed by his offspring being spared. And finally, a malingerer to the bitter end must place the lowest value on being conscripted.

On the other hand, after signaling madness, the madman is indifferent among all the outcomes that may be reached in the course of the game. Only from deceit can he end up winning nothing. For his part, Palamedes prefers above all the outcomes that for the malingerer are the most deeply wounding. He would prefer to conscript a madman (the berserker effect?) than to renounce a malingerer.

8.3. Behavioral–Strategic Analysis

In principle, it would be completely possible to introduce in place of the game tree of Figure 8.1 a normal form representation. The row player in this case would be pulling the strings for two marionettes: Odysseus the madman and Odysseus the malingerer. The column player would have the task of bringing Palamedes into the game.[3]

We obtain for each string-puller sixteen pure strategies, which recommend a particular course of action for each decision node. Thus, for example, the pure strategy m-m-o-s leaves the two Odysseuses to give the signal "madman" in the first stage of the game. Thereafter, the madman should offer up Telemachus, while the malingerer should spare him. One also says that the two forms of Odysseus (according to their signaling behavior) form a *pool* in the first stage but in the second stage give out *separating* signals.

In Figure 8.2 we have determined (based on the reduced puppeteer normal form)[4] for each configuration of pure strategies[5] the correspondingly weighted utility values (in accordance with the previously given a priori distribution).

Of course, the matrix in Figure 8.2 could be checked for the existence of equilibria (in mixed strategies). This, however, would be a case of gross malpractice. Since the game tree in Figure 8.1 indicates

---

[3] This technical trick occurs in the present case in complete concordance with the mythical course of events. Thus Hermes—the god of commerce, cunning, and theft—took the fate of his great-grandson Odysseus into his hands. Weighing in on the side of Palamedes was Pallas Athena—goddess of wisdom and warfare—also no slouch as a puppeteer.

[4] Whose rows and columns are commanded by Trojan warriors, courtesy of the Erich Lessing Culture and Fine Arts Archives.

[5] Six for the row player and five for the column player.

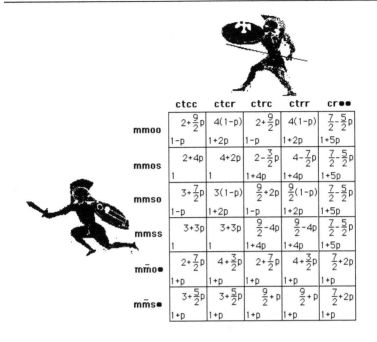

	ctcc	ctcr	ctrc	ctrr	cr••
mmoo	$2+\frac{9}{2}p$ / $1-p$	$4(1-p)$ / $1+2p$	$2+\frac{9}{2}p$ / $1-p$	$4(1-p)$ / $1+2p$	$\frac{7}{2}-\frac{5}{2}p$ / $1+5p$
mmos	$2+4p$ / 1	$4+2p$ / 1	$2-\frac{3}{2}p$ / $1+4p$	$4-\frac{7}{2}p$ / $1+4p$	$\frac{7}{2}-\frac{5}{2}p$ / $1+5p$
mmso	$3+\frac{7}{2}p$ / $1-p$	$3(1-p)$ / $1+2p$	$\frac{9}{2}+2p$ / $1-p$	$\frac{9}{2}(1-p)$ / $1+2p$	$\frac{7}{2}-\frac{5}{2}p$ / $1+5p$
mmss	$3+3p$ / 1	$3+3p$ / 1	$\frac{9}{2}-4p$ / $1+4p$	$\frac{9}{2}-4p$ / $1+4p$	$\frac{7}{2}-\frac{5}{2}p$ / $1+5p$
m̄m̄o•	$2+\frac{7}{2}p$ / $1+p$	$4+\frac{3}{2}p$ / $1+p$	$2+\frac{7}{2}p$ / $1+p$	$4+\frac{3}{2}p$ / $1+p$	$\frac{7}{2}+2p$ / $1+p$
mm̄s•	$3+\frac{5}{2}p$ / $1+p$	$3+\frac{5}{2}p$ / $1+p$	$\frac{9}{2}+p$ / $1+p$	$\frac{9}{2}+p$ / $1+p$	$\frac{7}{2}+2p$ / $1+p$

Figure 8.2. The (reduced) puppeteer normal form

an extensive game with perfect memory, one should rather fall back on the method of behavioral–strategic analysis.

In Figure 8.3 we have given an example of this procedure. After the Moerae, at the root of the game, have selected the type of Odysseus (in accordance with the game's a priori distribution $(1-p, p)$), the madman gives with probability 1, the malingerer with the (as yet unknown) probability α, the signal "madman."

In dependence on this (presumed) behavior, Palamedes will conjecture that if it is his turn to move in the information set "no signal," then he will find himself (with probability 1) in the right decision node. Now his best reply for this situation is to conscript Odysseus (the malingerer) with probability 1.

However, what occurs in the information set "madness is the signal"? The probability that it will be reached is $\alpha p + (1-p)$. As

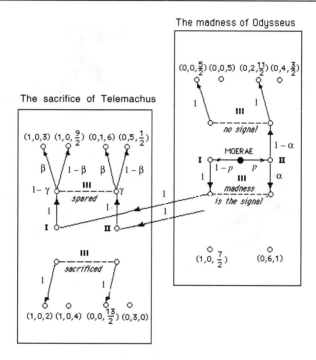

$(\alpha, 1-\alpha)$ ·· Behavioral strategy for the malingerer
at the first stage

$(\beta, 1-\beta)$ ·· Behavioral strategy of Palamedes
in the information set "spared"

$(\gamma, 1-\gamma)$ ·· A posteriori distribution of the two Odysseuses
in the information set "spared"

Figure 8.3. The art of behavioral strategic analysis

an a posteriori distribution of types we can deduce $(1-\gamma, \gamma)$ with

$$\gamma = \frac{\alpha p}{\alpha p + (1-p)}.$$

This distribution would hold as well for the information set "spared" if we were to assume that Palamedes decides on playing the Telemachus card with probability 1 and if both types (in the pool) spare Telemachus with the same probability 1.

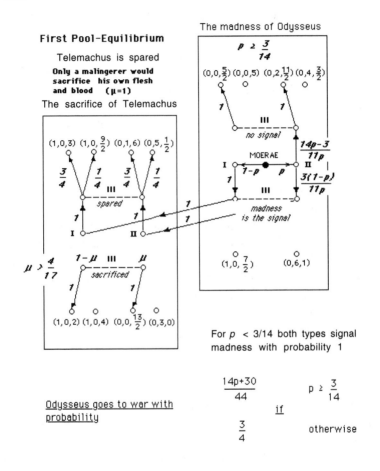

Figure 8.4. Telemachus is spared

How, finally, will Palamedes react in the set "spared"? Oddly enough, from the behavior of the malingerer we can determine Palamedes' unknown behavioral strategy that lets him choose with probability β the conscription orders for Odysseus. Since Odysseus the malingerer is indifferent between the signals from which he may choose, the following utility equation must hold:[6] $\beta + 5(1 - \beta) = 2$. Accordingly, one obtains for β the value $\frac{3}{4}$.

From the indifference of Palamedes in the set "spared" one can finally calculate the value of α. For this, one makes use of the a posteriori distribution $(1 - \gamma, \gamma)$. If Odysseus is conscripted, then Palamedes achieves the utility $3(1 - \gamma) + 6\gamma$. In the case of renunciation, on the other hand, he obtains $9(1 - \gamma)/2 + \gamma/2$. If these values agree, then

$$\alpha = \frac{3(1 - p)}{11p}.$$

Finally, let us turn our attention to the information set "sacrificed." If the players use the strategies described so far, this information set would not be reached at all. One might now be of the opinion that the behavior of Palamedes in this set plays no role whatsoever. What a false conclusion that would be! Namely, if Palamedes were to renounce his claim on Odysseus, then the malingerer could obtain a higher utility by a change of signal. The equilibrium in behavioral strategies depicted in Figure 8.4 would be thereby destroyed.

However, it will be maintained by a conjecture that Palamedes makes to issue conscription orders in the set "sacrificed."[7] This could, for example, go as follows: Only a malingerer would sacrifice his own flesh and blood.[8]

In any event, in the myth Telemachus[9] is spared. Faced with the choice of sacrificing his own flesh and blood, Odysseus admits defeat and is unmasked as a malingerer. The legend gives us a clear interpretation of Palamedes' conjecture: Only a madman would sacrifice Telemachus. We now know how dangerous and destabilizing this conjecture could be. It could lead straight to the equilibrium in Figure 8.5.

---

[6]On the left-hand side of this equation are the utilities that would accrue to the malingerer were he to give the signal "malingerer." On the right-hand side are the utilities that he would otherwise attain.

[7]The affair of conscription will later cost Palamedes dearly. He will be—through the cunning treachery of Odysseus—falsely accused of theft and executed.

[8]In Figure 8.4 we have determined further stabilizing conjectures off the path of equilibrium by means of the inequality $\mu > 4/17$.

[9]The name Telemachus in itself indicates the two options that present themselves to Odysseus. His name can be translated either as "one who remains far from battle" or as "one who battles afar."

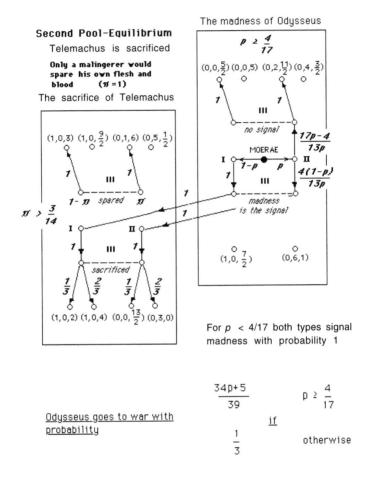

Figure 8.5. Telemachus is sacrificed

A Postlude in Rhyme

The game is winding down, and one feels the necessity to express fully all that one has not managed thus far to express. Has one perhaps not forgotten one or another important piece of advice, does one not wish at least to provide the knowledge-hungry wanderer one or another pointer to sources of wisdom or perhaps to an internet will o' the wisp?

The verdict as to success or failure is long since in. Seated at a dusty desk in a distant university town an academic executioner—one overlooked in the acknowledgments, perhaps, or whose works were not cited in the footnotes—stands in judgment over us. He has already sharpened his nib and is preparing to commit his verdict to writing (only in the rarest of cases will he do so in verse,[10] as happened to the unfortunate authors Drew Fudenberg and Jean Tirole in the case of their masterpiece [**30**])[11]

---

[10]As much as I with heavy heart must take full responsibility for the satirical verse that appears below, I have no choice but to assign a few quanta of responsibility to Steven Brams, who has improved my unspeakable English. Translator's note: In case the author's footnote has not elucidated the situation completely, namely, that the author wrote this verse in English, the translator wishes at this point to disavow any responsibility for such infernal nonsense.

[11]This unusual book review can be found at http://www.eos.tuwien.ac.at/OR/Mehlmann/Andis/Game4/game4.html.

The Mad Reviewer's Song[12]

He thought he saw a Hecatomb
    Of textbooks just in press:
He looked again, and found it was
    So Startling a Success.
"Each copy, sold at student price,
    Means earning (more or less)!"

He thought he saw a Book of Height
    (In inches) almost twelve:
He looked again, and found it was
    An MIT Press Elf.
"I won't be able, Lord," he said,
    "To put it on my shelf!"
He thought he heard a Wise Advice
    On publishing for free:
He heard again, and found it was
    An Unknown Referee.
"If you reject my work," he said,
    "Ken Binmore pays the fee!"

He thought he saw a Chapter on
    A differential game:
He looked again, and found it was
    A Long Prevailing Shame.
"A lot of reference," he said,
    "But what's about my name?"

He thought he saw a Strategy
    Undominated, strict:
He looked again, and found it was
    Quite Easy to Depict.
"I'll never play a game," he said,
    "So simple to predict!"

He thought he saw a Nash Profile
    Remaining unrefined:

---

[12] By Alexander Mehlmann with apologies to Lewis Carroll.

He looked again, and found it was
  Induction from Behind.
"Before more doubts arise," he said,
  "Apply it! Never mind!"

He thought he saw a Concept that
  One loves to understand:
He looked again, and found it was
  R. Selten's Trembling Hand.
"If Bonn's okay, we stay," he said,
  "Why point at Samarkand?"

He thought he saw a Dirty Face
  In common knowledge rage:
He looked again, and found it was
  A Game Without a Sage.
"Just read this book," he faintly said,
  While blushing at each page!

Envoi:

He thought he saw a Book Review
  With pages more than eight:
He looked again, and found it was
  A Really Awful Fate.
"If Fudenberg goes to Tirol(e),
  They wouldn't correlate!"

Appendix: Games in the Network of Networks

How long will the labors of our days find themselves pressed between the covers of a book? The new medium of the internet has shown itself to be a volatile but lightning-fast adversary to the printing press. The pros and cons are clearly laid out on the computer monitor. Information that one might once have managed to accumulate only after a lengthy correspondence today is only a microsecond away. This first of all diminishes the status of academic journals to that of a refereed archive, though on the other hand, it promotes the seepage of partially unfiltered and hasty results into the circulation of scientific discourse. Yet what are all these reservations against the possibility to obtain unfettered access to the source of game-theoretic knowledge? Thus, for example, the newest and best publications line up in battle formation to compete for the favor of the student of game theory:

Books on Game Theory Available on the World Wide Web

1. Michael Mesterton-Gibbons. *An Introduction to Game-Theoretic Modelling* [**62**]:
   http://www.math.fsu.edu/~mm-g/games.html

2. B. Nalebuff and A. Brandenburger. *Coopetition* [**66**]:
   http://mayet.som.yale.edu/coopetition/index2.html

3. D. Fudenberg and J. Tirole. *Game Theory* [**30**]:
   http://www-mitpress.mit.edu/book-home.tcl?isbn=0262061414

4. D. Fudenberg and D. Levine. *The Theory of Learning in Games* [**29**]:
   http://www-mitpress.mit.edu/book-home.tcl?isbn=0262061945

5. M. Osborne and A. Rubinstein. *A Course in Game Theory* [**72**]:
   http://www.chass.utoronto.ca/ osborne/cgt/index.html
   with corresponding solution manual:
   http://www.chass.utoronto.ca/~osborne/cgt/SOLS.HTM

6. Martin J. Osborne. *An Introduction to Game Theory* (to appear in 2000):
   http://www.chass.utoronto.ca/~osborne/igt/index.html

7. J. Weibull. *Evolutionary Game Theory* [**94**]:
   http://mitpress.mit.edu/book-home.tcl?isbn=0262731215

Yet all of this pales before the archive of research works on game theory at the University of Washington. This Ali Baba's treasure chamber offers to both sorcerers and their apprentices, if they can utter, or at least type in, the magic "open sesame" in the form http://econwpa.wustl.edu/months/game, the electronic pearls of wisdom, arranged chronologically, of more than forty "thieves." In place of Ali Baba's saddlebags the recommended apparatus for (down)loading and storing this treasure is a computer with access to the world wide web. Scientific journals are also to be extracted from the toils of the web. These and other game-theoretic sites are listed below.

Some Game Theory Treasures

1. *The International Journal of Game Theory.*
   http://www.tau.ac.il/ijgt/index.html

2. The journal *GAMES and Economic Behavior.*
   http://www.apnet.com/www/journal/ga.htm

3. *The Interuniversity Centre for Game Theory and Applications* at the University of Genoa.
   http://fismat.dima.unige.it/citg/citg.htm

4. The *Center For Rationality and Interactive Decision Theory* at the Hebrew University in Jerusalem.
   http://www.ma.huji.ac.il/~ranb/

5. The *International Society of Dynamic Games* at the University of Helsinki.
   http://www.hut.fi/HUT/Systems.Analysis/isdg/

6. *ELSE: The Centre for Economic Learning and Social Evolution.*
   http://else.econ.ucl.ac.uk/
   scientific home of Ken Binmore, author of the incomparable *Fun and Games* [**8**]

7. David Eppstein's page on combinatorial game theory.
   http://www.ics.uci.edu/~eppstein/cgt/

8. Nigel Howard's page on drama theory and confrontation analysis.
   http://www.nhoward.demon.co.uk/drama.htm

9. Roger McCain's introduction to game theory.
   http://www.coba.drexel.edu/economics/mccain/game/game.html

10. Kenneth N. Prestwich's highly informative and interactive Evolution and Game Theory website.
    http://science.holycross.edu/departments/biology/kprestwi/
    behavior/ESS/

11. The Game Theory Society.
    http://cwis.kub.nl/ few5/center/gts/

12. The game theory software GAMBIT.
    http://masada.hss.caltech.edu/ gambit/Gambit.html

13. Alvin Roth's page on game theory and experimental economics.
    http://www.pitt.edu/~alroth/alroth.html

14. Some invaluable prisoner's dilemma resources
    http://www.lifl.fr/IPD/ipd.html
    http://www.xs4all.nl/~helfrich/prisoner/
    http://www.kasprzyk.demon.co.uk/www/Dilemma.html

References

[1] AUMANN, R. J.: *Correlated Equilibrium as an Expression of Bayesian Rationality*. Econometrica, 55(1):1–18, 1987.

[2] AUMANN, R. J.: *What is Game Theory Trying to Accomplish?*. In ARROW, K. J. and S. HONKAPOHJA (eds.): *Frontiers of Economics*, 460–482. Blackwell, Oxford, 1987.

[3] AUMANN, R. J.: *Backward Induction and Common Knowledge of Rationality*. Games and Economic Behavior, 8:6–19, 1995.

[4] AUMANN, R. J. and A. BRANDENBURGER: *Epistemic Conditions for Nash Equilibrium*. Econometrica, 63(5):1161–1180, 1995.

[5] AXELROD, R.: *The Evolution of Cooperation*. Basic Books, New York, 1984.

[6] BERLEKAMP, E. R., J. H. CONWAY, and R. K. GUY: *Winning Ways for your Mathematical Plays: Games in General*, vol. 1. Academic Press, London, 1982.

[7] BERLEKAMP, E. R., J. H. CONWAY, and R. K. GUY: *Winning Ways for your Mathematical Plays: Games in Particular*, vol. 2. Academic Press, London, 1982.

[8] BINMORE, K.: *Fun and Games: A Text on Game Theory*. D. C. Heath and Company, Lexington, Massachusetts, 1992.

[9] BINMORE, K.: *Rationality and Backward Induction*. Technical report, Economics Department, University College London, Gower Street, London WC1E 6BT, UK, 1995.

[10] BOREL, E.: *La théorie du jeu et les équations intégrales à noyau symétrique*. Comptes Rendus de l'Académie des Sciences, 173:1304–1308, 1921.

[11] BOREL, E.: *Applications aux jeux d'hasard*. In *Traité du calcul des probabilités et de ses applications*. Gauthier-Villars, Paris, 1938.

[12] BRAMS, S. J. and W. DONALD: *Nonmyopic Equilibria in* 2×2 *Games*. Conflict Management and Peace Science, 6(1):39–62, 1981.

[13] BRAMS, S. J. and A. D. TAYLOR: *An Envy-Free Cake Division Protocol*. American Mathematical Monthly, 102(1):9–18, January 1995.

[14] BRAMS, S. J. and A. D. TAYLOR: *Fair Division: From Cake-Cutting to Dispute Resolution*. Cambridge University Press, Cambridge, 1996.

[15] BRAMS, S. J. and A. D. TAYLOR: *The Win–Win Solution: Guaranteeing Fair Shares to Everybody*, W. W. Norton & Company, 1999.

[16] CHO, I.-K. and D. M. KREPS: *Signaling Games and Stable Equilibria*. Quarterly Journal of Economics, CII(2):179–221, 1987.

[17] CONAN DOYLE, A.: *The Sign of the Four*. Oxford University Press, 1994.

[18] CONAN DOYLE, A.: *The Memoirs of Sherlock Holmes*. Penguin Books, 1970.

[19] DAWID, H. and A. MEHLMANN: *Two Population Contests and the Language of Genetics*. In PĂUN, G. (ed.): *Mathematical Linguistics and Related Topics*, 88–94. Editura Academiei Române, Bucureşti, România, 1995.

[20] DAWID, H. and A. MEHLMANN: *Genetic Learning in Strategic Form Games*. Complexity, 1(5):51–59, 1996.

[21] DAWKINS, R.: *The Selfish Gene*. Oxford, 1989.

[22] DIMAND, M. A. and R. W. DIMAND: *The History of Game Theory, Volume I: From the beginnings to 1945*. Routledge, London, 1996.

[23] DRESHER, M.: *The Mathematics of Games of Strategy*. Dover Publications, New York, 1981.

[24] DROSDEK, A.: *Sunzi und die Kunst des Krieges für Manager*. Langen Müller, München, 1996.

[25] EKELAND, I.: *The Broken Dice*. The University of Chicago Press, Chicago, 1993.

[26] ELIADE, M.: *Mythos und Wirklichkeit*. Insel Verlag, Frankfurt am Main, 1988.

[27] FEICHTINGER, G. and R. F. HARTL: *Optimale Kontrolle ökonomischer Prozesse*. Walter de Gruyter, Berlin, 1986.

[28] FEICHTINGER, G. and A. MEHLMANN: *Planning the Unusual: Applications of Control Theory to Nonstandard Problems*. Acta Applicandæ Mathematicæ, 7:79–102, 1986.

[29] FUDENBERG, D. and D. K. LEVINE: *Theory of Learning in Games*. MIT Press, 1998.

[30] FUDENBERG, D. and J. TIROLE: *Game Theory*. MIT Press, Cambridge, Massachusetts, 1991.

[31] GARDNER, M.: *New Mathematical Puzzles and Diversions*. Penguin, London, 1966.

[32] GARDNER, M.: *The Unexpected Hanging and Other Mathematical Diversions*. University of Chicago Press, 1991.

[33] GARDNER, M.: *Puzzles from Other Worlds*. Vintage, 1984.

[34] GLAZER, J. and C.-T. A. MA: *Efficient Allocation of a "Prize" : King Solomon's Dilemma*. Games and Economic Behavior, 1:222–233, 1989.

[35] GÜTH, W., R. SCHMITTBERGER, and B. SCHWARZE: *An Experimental Analysis of Ultimatum Bargaining*. Journal of Economic Behavior and Organization, 3:367–388, 1982.

[36] HARSANYI, J. C.: *Games with Incomplete Information Played by "Bayesian" Players, Part I*. Management Science, 14:159–182, 1967.

[37] HARSANYI, J. C.: *Games with Incomplete Information Played by "Bayesian" Players, Part II*. Management Science, 15:320–334, 1968.

[38] HARSANYI, J. C.: *Games with Incomplete Information Played by "Bayesian" Players, Part III*. Management Science, 15:486–502, 1968.

[39] HARSANYI, J. C. and R. SELTEN: *A General Theory of Equilibrium Selection in Games*. MIT Press, Cambridge, 1988.

[40] HESSE, H.: *Das Glasperlenspiel*. Suhrkamp Taschenbuch Verlag, Frankfurt am Main, 1972.

[41] HUGHES, P. and G. BRECHT: *Vicious Circles and Infinity: A Panoply of Paradoxes*. Jonathan Cape, London, 1975.

[42] HUYGENS, C.: *De ratiociniis in ludo aleæ*. In *Oeuvres complètes*, vol. 5, 35–47. La Haye, 1925.

[43] HYGINUS: *Fabula XCV*. In MARSHAL, P. (ed.): *Hygini Fabulæ*, Bibliotheca scriptorum Græcorum et Romanorum Teubneriana. Teubner, Stuttgart, 1988.

[44] ISAACS, R.: *Differential Games*. John Wiley & Sons, New York, 1965.

[45] KAHN, H.: *On Thermonuclear War*. Princeton University Press, Princeton, New Jersey, 1960.

[46] KAHN, H.: *On Escalation: Metaphors and Scenarios*. Praeger, New York, 1965.

[47] KARLIN, S.: *Mathematical Methods and Theory in Games, Programming and Economics*. Dover Publications, New York, 1992.

[48] KILGOUR, D. M.: *The Sequential Truel*. International Journal of Game Theory, 4:151–174, 1975.

[49] KILGOUR, D. M.: *Equilibria for Far-Sighted Players*. Theory and Decision, 16(2):135–157, 1984.

[50] KNUTH, D. E.: *Triel: A New Solution*. Journal of Recreational Mathematics, 6(1):1–7, 1973.

[51] KREPS, D. M.: *Game Theory and Economic Modelling*. Oxford University Press, New York, 1990.

[52] KREPS, D. M.: *A Course in Microeconomic Theory*. Harvester Wheatsheaf, New York, 1990.

[53] KREPS, D. M. and R. WILSON: *Sequential Equilibria*. Econometrica, 50:863–894, 1982.

[54] KUHN, H. W.: *Extensive Games and the Problem of Information*. In KUHN, H. W. and A. W. TUCKER (eds.): *Contributions to the Theory of Games*, vol. 2, 193–216. Princeton University Press, Princeton, 1953.

[55] LESSING, E.: *Die Odyssee: Homers Epos in Bildern erzählt*. Herder, Freiburg, 1966.

[56] LITTLEWOOD, J. E.: *A Mathematician's Miscellany*. Methuen and Company, London, 1953.

[57] MACKIE-MASON, J. K. and H. VARIAN: *Some economics of the internet*. Technical report, University of Michigan, 1993.

[58] MAYNARD SMITH, J.: *Evolution and the Theory of Games*. Cambridge University Press, Cambridge, 1982.

[59] MEHLMANN, A.: *Applied Differential Games*. Plenum Press, New York, 1988.

[60] MEHLMANN, A.: *De Salvatione Fausti: Die Wette zwischen Faust und Mephisto im Lichte von spieltheoretischem Calcül und neuerem Operational Research*. no. 5 in *Litzelstetter Libellen*. Ekkehard Faude Verlag, Konstanz, 1989.

[61] MEHLMANN, A. and R. WILLING: *Eine spieltheoretische Analyse des Faustmotivs*. Mathematische Operationsforschung und Statistik, 15(2):243–252, 1984.

[62] MESTERTON-GIBBONS, M.: *An Introduction to Game-Theoretic Modelling*. Addison-Wesley, Redwood City, 1992.

[63] MÉZIRIAC, B. DE: *Problèmes plaisants et délectables, qui se font par les nombres*. Lyon, 1612.

[64] MOORE, E. H.: *A generalization of the game called Nim*. Annals of Mathematics, 11:93–94, 1909.

[65] MOSES, Y., D. DOLEV, and J. Y. HALPERN: *Cheating Husbands and Other Stories: A Case Study of Knowledge, Action and Communication*. Distributed Computing, 1:167–176, 1986.

[66] NALEBUFF, B. J. and A. M. BRANDENBURGER: *Coopetition— kooperativ konkurrieren: Mit der Spieltheorie zum Unternehmenserfolg*. Campus Verlag, Frankfurt, 1996.

[67] NASH, J. F.: *The Bargaining Problem*. Econometrica, 18:155–162, 1950.

[68] NASH, J. F.: *Equilibrium Points in n-Person Games*. Proc. Nat. Acad. Sci. U.S.A., 36:48–49, 1950.

[69] NASH, J. F.: *Non-Cooperative Games*. Annals of Mathematics, 54:286–295, 1951.

[70] NEUMANN, J. VON: *Zur Theorie der Gesellschaftsspiele*. Mathematische Annalen, 100:295–300, 1928.

[71] NEUMANN, J. VON and O. MORGENSTERN: *Theory of Games and Economic Behavior*. Princeton University Press, Princeton, 1944.

[72] OSBORNE, M. J. and A. RUBINSTEIN: *A Course in Game Theory*. MIT Press, Cambridge, Massachusetts, 1994.

[73] POUNDSTONE, W.: *Prisonner's Dilemma: John von Neumann, Game Theory, and the Puzzle of the Bomb*. Doubleday, New York, 1992.

[74] RAPOPORT, A.: *The Use and the Misuse of Game Theory*, in *Mathematical Thinking in Behavioral Sciences*. W. H. Freeman and Co., San Francisco, 1968.

[75] RAPOPORT, A. and A. M. CHAMMAH: *Prisonner's Dilemma: A Study in Conflict and Cooperation*. University of Michigan Press, Ann Arbor, second edition, 1970.

[76] REZZORI, G. VON: *Die schönsten maghrebinischen Geschichten*. Rowohlt Verlag, Hamburg, 1953.

[77] REZZORI, G. VON: *Maghrebinische Geschichten*. Rowohlt Taschenbuch Verlag, 1994. (Translated into English by Catherine Hutter as *Tales of Maghrebinia* New York, Harcourt, Brace & World, 1962.

[78] ROTH, A. E., V. PRASNIKAR, M. OKUNO-FUJIWARA, and S. ZAMIR: *Bargaining and Market Behavior in Jerusalem, Ljubljana, Pittsburgh, and Tokyo: An Experimental Study*. American Economic Review, 81(5):1068–1095, 1991.

[79] ROSENTHAL, R.: *Games of Perfect Information, Predatory Pricing, and the Chain-Store Paradox*. Journal of Economic Theory, 25:92–100, 1981.

[80] RUBINSTEIN, A.: *The Electronic Mail Game: Strategic Behavior under "Almost Common Knowledge."* The American Economic Review, 79(3):385–391, 1989.

[81] SAMET, D.: *Hypothetical Knowledge and Games with Perfect Information*. Games and Economic Behavior, 17(2):230–251, 1996.

[82] SAMET, D.: *Counterfactuals in Wonderland*. Faculty of Management, Tel Aviv University, Tel Aviv, Israel, 1997.

[83] SAMET, D.: *Rationality, Counterfactuals and No-matter-what Theories*. Faculty of Management, Tel Aviv University, Tel Aviv, Israel, 1997.

[84] SELTEN, R.: *Spieltheoretische Behandlung eines Oligopolmodells mit Nachfrageträgheit*. Zeitschrift fur die gesamte Staatswissenschaft, 121:301–324, 667–689, 1965.

[85] SELTEN, R.: *Reexamination of the Perfectness Concept for Equilibrium Points in Extensive Games*. International Journal of Game Theory, 4:25–55, 1975.

[86] SELTEN, R.: *The Chain-Store Paradox*. Theory and Decision, 9:127–159, 1978.

[87] SHUBIK, M.: *Does the Fittest Necessarily Survive?*. In SHUBIK, M. (ed.): *Readings in Game Theory and Political Behavior*, 43–46. Doubleday, Garden City, N. Y., 1954.

[88] SHUBIK, M.: *Game Theory in the Social Sciences: Concepts and Solutions*. MIT Press, Cambridge, Massachusetts, 1982.

[89] STEINHAUS, H.: *Mathematical Snapshots*. Oxford University Press, New York, 1969.

[90] THÉPOT, J.: *Irréversibilité et Décision Economique selon Gustave Flaubert*. Revue d'Economie Politique, 91:494–498, 1981.

[91] THUCYDIDES: *The Peloponnesian War*. Indianapolis, Hackett, 1998.

[92] TURNER, PAUL E. AND CHAO, LIN: *Prisoner's dilemma in an RNA virus*. Nature, 398:441–443, 1999.

[93] WALDEGRAVE, J.: *Minimax solution to a 2-person zero-sum game, reported 1713 in letter from P. de Montmort to N. Bernoulli*. In BAUMOL, W. J. and S. GOLDFIELD (eds.): *Precursors of Mathematical Economics*, 3–9. London School of Economics, London, 1968.

[94] WEIBULL, J. W.: *Evolutionary Game Theory*. The MIT Press, Cambridge, Massachusetts, 1995.

[95] WILLIAMS, J. D.: *The Compleat Strategyst: Being a Primer on the Theory of Games of Strategy*. Dover Publications, New York, 1986.

[96] ZERMELO, E.: *Über eine Anwendung der Mengenlehre auf die Theorie des Schachspiels*. In HOBSON, E. W. and A. E. H. LOVE (eds.): *Proceedings of the Fifth International Congress of Mathematicians*, 501–504, Cambridge, 1913. Cambridge University Press.

Index